Sense and Nonsense

Sense and Nonsense

Evolutionary Perspectives on Human Behaviour

THIRD EDITION

Gillian R. Brown
Kevin N. Lala

OXFORD
UNIVERSITY PRESS

Great Clarendon Street, Oxford, OX2 6DP,
United Kingdom

Oxford University Press is a department of the University of Oxford.
It furthers the University's objective of excellence in research, scholarship,
and education by publishing worldwide. Oxford is a registered trade mark of
Oxford University Press in the UK and in certain other countries

© Gillian R. Brown and Kevin N. Lala 2024

The moral rights of the authors have been asserted

First Edition published 2002
Second Edition published 2011
Third Edition published 2024

All rights reserved. No part of this publication may be reproduced, stored in a retrieval system, transmitted, used for text and data mining, or used for training artificial intelligence, in any form or by any means, without the prior permission in writing of Oxford University Press, or as expressly permitted by law, by license or under terms agreed with the appropriate reprographics rights organization. Inquiries concerning reproduction outside the scope of the above should be sent to the Rights Department, Oxford University Press, at the address above.

You must not circulate this work in any other form
and you must impose this same condition on any acquirer

Published in the United States of America by Oxford University Press
198 Madison Avenue, New York, NY 10016, United States of America

British Library Cataloguing in Publication Data
Data available

Library of Congress Control Number: 2024944418

ISBN 978–0–19–890820–3

DOI: 10.1093/oso/9780198908203.001.0001

Printed and bound by
CPI Group (UK) Ltd, Croydon, CR0 4YY

Oxford University Press makes no representation, express or implied, that the drug dosages in this book are correct. Readers must therefore always check the product information and clinical procedures with the most up-to-date published product information and data sheets provided by the manufacturers and the most recent codes of conduct and safety regulations. The authors and the publishers do not accept responsibility or legal liability for any errors in the text or for the misuse or misapplication of material in this work. Except where otherwise stated, drug dosages and recommendations are for the non-pregnant adult who is not breast-feeding

Preface to third edition

The first edition of *Sense and Nonsense: Evolutionary Perspectives on Human Behaviour* was published in 2002, and the second edition in 2011. Sufficient time had now passed that we decided to revise the content to reflect advances in the field and to evaluate whether any of the points of contention had been resolved. Our aim with this edition, as with the previous ones, has been to provide an accurate and, as far as possible, impartial representation of the contemporary sub-fields of human evolutionary behavioural sciences that have their roots in the 'sociobiology debates' of the 1970s and 1980s. These original debates surrounded the question of whether ideas from evolutionary biology could be applied to human behaviour and, if so, how should they be applied. Would the methods need modifying for our own species, given our clear reliance on culture? We believe that the field is closer to answering these questions but that continued interdisciplinary research will be required, drawing from many fields, including anthropology, archaeology, psychology, zoology, genetics, and mathematics. Such interdisciplinary work is difficult, in terms of requiring individual researchers to have a broad skill set or to build fruitful cross-disciplinary collaborations. Understanding the evolution of human behaviour was unlikely to be easy, but the costs of carrying out quick, superficial research in this area are potentially large, given that unsupported claims about human behaviour can have real-world implications that are often particularly harmful for those from disadvantaged or under-represented groups. We want to encourage the ethics of this area of research to be at the forefront of academics' minds.

What changes have been made in this new edition? The basic structure of the book remains the same. After a brief introductory chapter, we provide a condensed history of the application of evolutionary theory to human behaviour, starting with the writings of Charles Darwin and ending with the 'human ethologists'. The following chapter describes the core ideas of 'sociobiology' and the controversy surrounding their application to human behaviour. The next four chapters cover the main contemporary sub-fields of this discipline, namely human behavioural ecology, evolutionary psychology, cultural evolution, and gene–culture coevolution, as in the previous edition. For each chapter, we describe the key concepts underlying the relevant sub-field, present updated case studies, and provide a critical evaluation of the sub-field. All of these sub-fields have been active during this period, and we have documented advances throughout. For instance, human behavioural ecologists have applied their methods to ask applied questions, evolutionary psychologists have moved away from a focus on 'domain specificity', cultural evolution researchers have collected empirical datasets that are relevant to their mathematical models, and gene–culture coevolution researchers now have access to detailed genomic data from a range of species. The order of authorship for this edition of the book has switched, which reflects the fact that GRB took the lead in writing this revised edition. KNL's surname has changed from the earlier editions, as he has reverted back to his birth name, which reflects his Indian Parsi heritage.

The current social, political, and scientific climate in which Western science takes place has also changed since the first two editions of this book. Although we do not return to these points in detail, we want to acknowledge the impact of racial and gender inequalities, and histories of colonialism, on how science is undertaken and who gets to contribute to knowledge bases, plus the impact of the Black Lives Matter and #MeToo movements in bring these topics to the fore within society in general. In academia, we have witnessed the 'replication crisis' in psychology and a greater push towards 'open science', whereby scientific findings are made available beyond the walls of academia. The ways in which money can influence the direction of science have also become apparent, for instance, through the controversial funding of research by wealthy donors and through personal contacts with those of disrepute. The company kept by scientists impacts their academic reputations but not always their academic careers, as well as sometimes excluding those who have higher ethical thresholds for their work. We value transparency, honesty, and high ethical standards within science. We are extremely keen that the history of applying evolutionary theory to human behaviour (as briefly outlined in Chapter 2) is understood and appreciated, as the spectre of 'scientific racism', for example, remains an active threat. We touch briefly on this topic in Chapter 8, though a more in-depth evaluation would have been warranted.

We are grateful for visiting fellowships at the Konrad Lorenz Institute (KLI) for Evolution and Cognitive Research in Klosterneuburg, Austria. The KLI provided an intellectually stimulating and relaxing environment for some of the writing. We have also been supported by the University of St Andrews, UK, where we have worked since the early 2000s. Thank you to all the academic staff members and postgraduate students who have provided feedback on earlier draft chapters for this edition, including Clark Barrett, Sven

Kasser, Jeremy Koster, David Lawson, Bobby May, Siobhán Mattison, Richard McElreath, Alex Mesoudi, Csilla Pákozdy, Peter Richerson, Rebecca Sear, Tim Temizyürik, and Jinwen Xie, as well as those who provided feedback on the earlier editions. Thank you to the staff at Oxford University Press, particularly Martin Baum, for continued support. Finally, we are grateful to everyone who has used *Sense and Nonsense* when teaching their undergraduate and postgraduate students, as well as our general readers. We are aware that previous editions have been used on numerous university courses around the world, and we hope that this version is equally useful for guiding and educating students and researchers. Our overarching goal has been to help readers to navigate this field and decide for themselves whether we are collectively any closer to understanding the evolution of human behaviour.

G.R.B. and K.N.L.
St Andrews
March 2024

Preface to second edition

A year after the two hundredth anniversary of Charles Darwin's birth, evolutionary accounts of human behaviour are as prevalent as ever. Such accounts also remain remarkably contentious, still attracting legions of hostile criticism and exciting vigorous, impassioned defences. Whether you are an admirer or not, there can be no doubt that the field has retained a high profile, both within and outside academia, and that it remains remarkably fertile. Nonetheless, in the eight years since the first edition was completed, a great deal has changed. In many non-trivial respects, virtually all domains of the human evolutionary behavioural sciences have made significant progress. We are grateful to Oxford University Press for providing us with the opportunity to produce a second edition to recount these changes.

The last eight years have been quite an extraordinary period. For instance, the human genome has been sequenced, with dramatic implications for the human evolutionary behavioural sciences, as we discuss at some length in this book. Important advances have been made in our understanding of recent human evolution, which have affected the field from the outside. Significant developments have also occurred within the field, such as the broadening of evolutionary psychology methodology, the appearance of cultural evolution experimentation, and the application of phylogenetic approaches to culture. Unfortunately, the rapid progress in the field of cultural evolution has not been matched within the field of memetics. To reflect the changing literature, Chapter 6, which was previously devoted to memes, has been restyled as a cultural evolution chapter, with memes just making up a small part.

Our aim with *Sense and Nonsense* has always been to present a critical evaluation of the human evolutionary behavioural sciences, while at the same time endeavouring to be constructive, fair, and impartial. Others have chosen to be more aggressive and confrontational in their critiques. We do not believe that this kind of criticism is constructive, as it does not generally change the working practices of researchers within the field but merely promotes polarization among advocates and critics. In contrast, informed and broadly sympathetic critical analysis can potentially bring about change and promote a deeper understanding. Ultimately, we hope to have played a small part in building a rich, multifaceted, pluralistic, and disciplined field of evolution and human behaviour.

As with the first edition, our intended audience for the book is undergraduate and postgraduate students (e.g. in anthropology, psychology, zoology, and the human sciences), our colleagues in the field, and lay persons interested in evolutionary explanations of human behaviour. Numerous colleagues from around the world have informed us that they are using *Sense and Nonsense* on their taught courses, and we are very grateful for this support. We would also like to thank the following researchers for providing comments on the second edition, or helping in other respects: Robert Boyd, Tom Dickins, Marc Feldman, Dan Fessler, Russell Gray, Kim Hill, Jeremy Kendal, David Lawson, Alex Mesoudi, Mark Pagel, Luke Rendell, Rebecca Sear, and Amanda Seed. We are also grateful to Katherine Meacham for secretarial support and to the staff at Oxford University Press.

K.N.L. and G.R.B.
St Andrews
July 2010

Preface to first edition

Can evolutionary theory help us to understand human behaviour and society? Many evolutionary biologists, anthropologists, and psychologists are optimistic that evolutionary principles can be applied to human behaviour, and have offered evolutionary explanations for a wide range of human characteristics, such as homicide, religion, and sex differences in behaviour. Others are sceptical of these interpretations, and stress the effects of learning and culture. They maintain that human beings are too special to study as if they were just another animal—after all, we have complex culture, language, and writing, and we build houses and programme computers. Perhaps both of these stances are right to a degree. Some aspects of our behaviour may be more usefully investigated using the methods of evolutionary biology than others. The challenge for scientists will be to determine which facets of humanity are open to this kind of analysis, and to devise definitive tests of any hypotheses concerning our evolutionary legacy. For those of us fascinated by this challenge, knowledge of the diverse methods by which human behaviour is studied from an evolutionary perspective would seem a prerequisite. In this book, we outline five evolutionary approaches that have been used to investigate human behaviour and characterize their methodologies and assumptions. These approaches are sociobiology, human behavioural ecology, evolutionary psychology, memetics, and gene–culture coevolution. For each, we discuss their positive features and their limitations and in the final chapter we compare their relative merits.

Innumerable popular books have already been published that discuss human behaviour and evolution, e.g. *The Selfish Gene* (Dawkins, 1976), *The Rise and Fall of the Third Chimpanzee* (Diamond, 1991), *Darwin's Dangerous Idea* (Dennett, 1995), *How the Mind Works* (Pinker, 1997), and *The Meme Machine* (Blackmore, 1999). Each gives a unique and stimulating view of human nature. However, such books usually take a single viewpoint on human evolution, frequently identifying with a particular school, such as evolutionary psychology or memetics. There have also been academic books published from these different perspectives, such as *Culture and the Evolutionary Process* (Boyd and Richerson, 1985), *The Adapted Mind* (Barkow et al., 1992), *Adaptation and Human Behavior* (Cronk et al., 2000) and *Darwinizing*

Culture: The Status of Memetics as a Science (Aunger, 2000a). In contrast to these, our book takes a pluralistic approach, highlighting how different researchers have divergent views on the best way to use evolutionary theory to study humanity. Heated debates and personal attacks have often ensued. Some of the approaches described will be new to many readers, as the theories on which they are based have generally not made it further than the specialist scientific literature. In presenting these fields we endeavour to translate these methodologies into easily understandable examples, and thereby make accessible new perspectives on how human behaviour and culture can be interpreted.

In writing this book, we pursue three goals. First, like Eric Alden Smith and colleagues (Smith et al., 2001), we see a need for 'a guide for the perplexed' for those of us who have struggled to understand the plethora of confusing terms and apparent differences of opinion and approach in the use of evolutionary theory to study human behaviour. Secondly, in line with a long tradition of researchers based at the Sub-Department of Animal Behaviour at the University of Cambridge, where we work, we believe that research in this domain is best served by a rigorous, self-critical science, and that the study of behaviour requires a broad perspective that incorporates questions such as how behaviour develops over an individual's lifetime as well as questions about how behaviour evolves. Thirdly, we see great value in pluralism in the use of methodology, and the integration of approaches. We hope to have made a small contribution in each of these regards.

This book does not provide an overview of the use of evolutionary theory in areas such as economics, law, and literature. We acknowledge the important work in these areas, but would rather maintain the length of the book as it is, and remain within more familiar territory. To those whose research is addressed, we hope that a fair synopsis is provided and are very grateful to all of the experts who have taken the time to discuss their work with us. We have personal views on the relative merits of the five schools of thought described; however, we have attempted to treat each approach evenly by asking leading members in the fields to help us to present their views accurately. Perhaps our profiles of the alternative approaches will highlight to some researchers how the methods may be integrated in the future, as well as draw attention to the conflicts that are yet to be resolved. Of those who currently deny the relevance of biology to the study of human behaviour, we hope that we might perhaps make some converts. More realistically, we hope that their scepticism will be tempered by the realization that not all researchers in this area are genetic determinists, Panglossian adaptationists, or wanton biologizers, and that many are prepared to place emphasis on non-biological and even non-evolutionary explanations.

Our intention is that this introductory book will be of use to undergraduate and postgraduate students (e.g. in zoology, anthropology, and psychology) and to experts on one approach who would like to know more about the other perspectives, but also to lay persons interested in evolutionary explanations of human behaviour. We have tried to write the text so that anyone interested in this subject area will find the material easy to comprehend. Our intention is not to provide a textbook review of the whole subject area, but rather to give a taste of the various options. For readers who would like to know more about a particular perspective, further reading is provided at the end of the book.

The most enjoyable aspect of writing this book has been the opportunity to interact with many of the leading authorities in this area of research. We have been overwhelmed by the kindness and generosity of those who have discussed their work with us and have commented on chapters of the book: we have learned so much from them. We would like to thank the following people for commenting on one or more chapters and for discussing the material in the book: Robert Aunger, Pat Bateson, Gillian Bentley, Susan Blackmore, Monique Borgerhoff Mulder, Robert Boyd, Nicky Clayton, Tim Clutton-Brock, Leda Cosmides, Alan Costall, Nick Davies, Richard Dawkins, Daniel Dennett, Robin Dunbar, Dominic Dwyer, Marc Feldman, Dan Fessler, Jeff Galef, Oliver Goodenough, Russell Gray, Kristen Hawkes, Robert Hinde, Sarah Blaffer Hrdy, David Hull, Rufus Johnstone, Mark Kirkpatrick, Richard Lewontin, Elizabeth Lloyd, John Maynard Smith, John Odling-Smee, Sally Otto, Henry Plotkin, Peter Richerson, Eric Alden Smith, Elliott Sober, John Tooby, Markus Vinzent, and Ed Wilson. We are also particularly grateful to Jeffrey Brown, Dominic Dwyer, Robert Hinde, Claire Laland, Bob Levin, Ed Morrison, and John Odling-Smee for reading the entire book and providing detailed feedback. We would like to thank the members of the Discussion Group at Madingley (Roz Almond, Yfke van Bergen, James Curley, Rachel Day, Tim Fawcett, Will Hoppitt, Jeremy Kendal, Bob Levin, and Liz Pimley), who worked through early drafts of each chapter with us, and provided very valuable input and encouragement. We were helped by comments from Mat Anderson, Martin Daly, Jean Dobel, Richard McElreath, Heather Proctor, and Joan Silk. Thanks also to Martin Baum at OUP and to Sheila Watson of Watson Little Ltd for their advice and guidance. This research was supported by a Royal Society University Research Fellowship to KNL and Medical Research Council funding to GRB. Finally, we are grateful to Ed Wilson and

Sarah Blaffer Hrdy for their enthusiastic support and encouragement, the memories of which have kept us going when we thought that we might have bitten off more than we could chew.

K.N.L. and G.R.B.
Cambridge
March 2002

Contents

1. Sense and nonsense	1
2. A history of evolution and human behaviour	17
3. Human sociobiology	50
4. Human behavioural ecology	78
5. Evolutionary psychology	108
6. Cultural evolution	139
7. Gene–culture coevolution	169
8. Comparing and integrating approaches	199
References	*225*
Index	*259*

1
Sense and nonsense

The human species is unique. We contemplate why we are here, and we seek to understand why we behave in the way that we do. Among the most compelling answers that modern science can provide for these eternal questions are those based on evolutionary theory. Few ideas have excited more reflection than Charles Darwin's theory of evolution by natural selection. Currently, evolutionary thinking is everywhere. Up-and-coming young executives look to evolutionary lore for the latest in business acumen. Prisons use evolutionary logic to reduce tension among inmates. Medics exploit knowledge of human evolution to revise diagnoses and develop new treatments. Even grocery stores are taking on evolutionarily minded psychologists as consultants to tell them how best to stack their shelves.

Judging by its media profile and its representation in academic and popular science, evolutionary theory would seem to provide the solution to almost every puzzle. Every day, the newspapers abound with evolutionary explanations for human characteristics such as 'aggression' or 'criminal behaviour', while book shops are overflowing with popular science texts boldly asserting that evolution will reveal how to find your perfect partner, how to have a successful marriage, or how to make it to the top of your profession. We are told by various authors that human beings behave like 'naked apes' floundering in a modern world, that male promiscuity is inevitable, that our political orientations reflect ancient tribalism, and that everything we do is ultimately a means to propagate our genes. However, to what extent can human behaviour really be understood by taking an evolutionary viewpoint? What truth lies behind the newspaper reports and popular science stories? The aim of this book is to provide some answers to these questions.

Clearly, for many academic researchers, taking an evolutionary viewpoint is a fruitful means of interpreting human behaviour and society. Not only does evolution dominate the biological sciences, but it has also made inroads into the social sciences, with thriving disciplines such as 'evolutionary psychology', 'evolutionary anthropology', and 'evolutionary economics'. Yet if an evolutionary perspective is so productive, why isn't everyone using it? What is it that leads the majority of professional academics in the social sciences not only to ignore evolutionary methods, but in many cases to be extremely hostile to the arguments? If evolutionary theory is having ramifications that permeate every aspect of human society, it would be reassuring to have confidence in the claims made in its name. In which case, should we not be concerned that some of the world's leading evolutionary biologists are highly critical of the manner in which fellow academics employ evolution to shed light on human behaviour?

The reality is that evolutionary perspectives on human behaviour frequently incite controversy, even among the scientists themselves. Evolutionary theory is one of the most fertile, wide-ranging, and inspiring of all scientific ideas. It offers a battery of methods and hypotheses that can be used to interpret human behaviour. However, the legitimacy of this exercise is at the centre of a heated controversy that has raged for more than a century. Ultimately, the disquiet traces back to past misuses of evolutionary reasoning to bolster prejudiced ideas and ideologies. Although these transgressions often resulted from distortions of Darwinian thought, this darker side has resulted in many academic disciplines characterizing the use of evolution to elucidate humanity as harmful and even dangerous. Many researchers within the social sciences and humanities remain extremely uncomfortable with evolutionary approaches. Consequently, disputes over evolutionary interpretations of humanity have fostered a polarization of thought.

As evolutionary theory becomes more technical, many people find it difficult to distinguish basic biological truths from speculative stories or prejudicial argument. Like all areas of science, the work in this field varies greatly in quality. At its best, evolutionary analyses of human behaviour meet the highest standards, but at the other extreme we find a sensationalistic 'tabloid' pseudoscience. Zealous advocates of evolutionary analyses rarely highlight the limitations of their findings, while impassioned critics seldom acknowledge that there is some merit to an evolutionary analysis. This book outlines the most prominent evolutionary approaches and theories currently being used to study human behaviour, guiding the reader through the mire of confusing terminology, claim and counter-claim, and polemic statements. We will explore to what extent human behaviour can legitimately be studied using these evolutionary methods. At the same time, we will consider whether there are unique features of human society and culture that sometimes render such methods impotent. Both evolutionary arguments and the allegations of the critics will be subjected to careful scrutiny. By the end of the book, the reader should feel better placed to assess the legitimacy of claims made about human behaviour under the name of evolution.

Taking the middle ground

An example of the controversy that can surround the use of evolution to interpret human behaviour is provided by the extraordinary response to an academic textbook written by Edward O. Wilson, an eminent Harvard University professor. In 1975, Wilson produced an encyclopaedic book on animal behaviour entitled *Sociobiology: The New Synthesis* (Wilson, 1975a). Although textbooks on animal behaviour rarely become bestsellers or arouse much media attention under normal circumstances, Wilson's tome was different. In the final chapter of the book, Wilson described how the latest advances in the study of animal behaviour, particularly the insights of biologists Robert Trivers and William Hamilton, might explain many aspects of human behaviour. He provided biological explanations for a broad array of controversial topics, including human aggression, religion, homosexuality, xenophobia, and differences between the sexes. He also predicted that it would not be long before the social sciences were subsumed within the biological sciences.

Wilson's book provoked an uproar and launched what is now known as the 'sociobiology debate', which raged throughout the 1970s and 1980s. Social scientists bitterly disputed Wilson's claims, found fault with his methods, and dismissed his explanations as speculative stories. Intriguingly, among the most prominent critics were two members of Wilson's own department at Harvard—evolutionary biologists Richard Lewontin and Stephen J. Gould—who vehemently attacked the book in the popular press as simple-minded and reductionist. Yet most biologists could see the potential of the sociobiological viewpoint, which had paid great dividends in understanding non-human animals, and many were drawn into using these new tools to interpret humanity. The debate became polarized and highly political, with the sociobiologists accused of bolstering right-wing conservative values and the critics associated with Marxist ideology (for more on this debate, see Chapter 3). In the midst of this controversy, when emotions were raised and knee-jerk reactions common, the position of John Maynard Smith, one of the world's leading evolutionary biologists at the time, stood out for its balanced judgement and fairness. In the heat of the debate, Maynard Smith retained a dignified intermediate position, supporting science over politics. He remained very conscious of the dangers of an inappropriate use of biology, at the same time as being angry at much of the unjust criticism directed at Wilson. In an interview in 1981, he stated:

> I have a lot of the gut feelings of my age of being horrified and scared of the application of biology to the social sciences—I can see ... race theories, Nazism, anti-semitism and the whole of that. So that my initial gut reaction to Wilson's *Sociobiology* was one of considerable annoyance and distress. (1981; quoted in Segerstråle, 2000, pp. 240–241)

Yet in a balanced review of *Sociobiology* (Maynard Smith, 1975), he described the book as making 'a major contribution' to an understanding of animal behaviour and was careful to stress its many positive features.

In her analysis of the sociobiology debate, sociologist Ullica Segerstråle (2000) stated that few scientists were well positioned to be communicators or 'arbiters' between the sociobiologists and their critics, because few scientists understood both sides. In addition to Maynard Smith, Segerstråle (2000) singled out British ethologist Patrick Bateson for taking the middle ground and playing a mediating role between the protagonists. The debate had become so polarized that Maynard Smith later admitted:

> I find that if I talk to Dick Lewontin or Steve Gould for an hour or two, I become a real sociobiologist, and if I talk to someone like Wilson or Trivers for an hour or two, I become wildly hostile to it. (1981; quoted in Segerstråle, 2000, p. 241)

In this book, we endeavour to follow Maynard Smith's lead and take the middle ground between the positions of advocates of evolutionary approaches to the study of human behaviour and their critics. We hope that we have also provided a balanced, central view that outlines the positive features of evolutionary methods but does not shy away from stating where we find the arguments suspect and remains vigilant to the dangers of irresponsible biologizing. Some researchers appear to believe that all aspects of behaviour are best described by reference to human evolutionary history. We do not take this line and believe that alternative explanations of human behaviour must always be considered.

The high temperature of the sociobiology debate, and the severity of the criticism, appeared to engender a 'circle the wagons' mind set among many human sociobiologists. When the flak was heavy they closed ranks, put up a united front, and some tacitly agreed not to criticize each other's work openly for fear of providing ammunition for the opposition. At the founding meeting of the Human Behavior and Evolution Society (HBES) in Evanston, Illinois, USA, in 1989, the society's President, William Hamilton, gave an address in which he described scholars interested in the evolutionary basis of human behaviour as 'a small, besieged group' (Segerstråle, 2000). Some people present at the time recall Hamilton urging enthusiasts not to worry if their theories were crazy or their hypotheses untestable but to march ahead boldly without fear of the consequences. One leading researcher, who was then a junior member of the society, recalls voicing the concern that this message would inadvertently foster a less rigorous approach to science, but this view received little support at the time, and overt criticism of other's work was actively discouraged. We would not wish to stifle creativity within this field of academic research, and we recognize that there is a time for, and value to, brainstorming and speculation. Nonetheless, we believe that any scientific field needs to evaluate its own assumptions and research methods to progress. Now that research into human behaviour and evolution is well established, the strongest defence against external criticism would be to maintain the highest standards of science.

Within the broad community of researchers who take an evolutionary approach to investigate human behaviour, some individuals would appear to identify with particular sub-fields and see important distinctions between the approach of their sub-field and that of the alternatives. Others recognize no 'factions' and see no major differences in approach between the leading 'schools'. As the former position would still appear to represent a large proportion of the field, in this book, we characterize five different approaches to the study of human behaviour that have emerged since some key conceptual advances in the 1970s. These five approaches are 'human sociobiology', 'human behavioural ecology', 'evolutionary psychology', 'cultural evolution', and 'gene–culture coevolution'. We believe that the theory and methods of these sub-fields differ in important ways, and we have emphasized these distinctions. Some of these differences may stem from having their roots in different research traditions and academic disciplines, while others are more ideological. In the final chapter of this book, we compare evolutionary perspectives in an attempt to isolate which techniques are legitimate and insightful, and those that are found wanting. We go on to consider how the best of each approach could be integrated into a coherent field of evolutionary human behavioural sciences and discuss where the most contentious points of disagreement remain.

A guide for the bewildered

To the outsider, and even to many on the inside, the field of human behaviour and evolution is riddled with confusing terminology. There are 'Darwinian psychologists', 'evolutionary anthropologists', 'cultural evolutionists', and 'gene–culture coevolutionists'. There is 'evolutionary psychology', 'dual-inheritance theory', 'human behavioural ecology', and 'memetics'. Some people cast all these approaches as 'human sociobiology', whereas others are at pains to distinguish between them, or even change their own labels over time. For instance, Britain's most famous 'sociobiologist', Richard Dawkins, described himself as an 'ethologist' in his books and writings up until 1985, when he switched to classifying himself as a sociobiologist to counteract Wilson's increasingly vocal critics. In the millennium edition of *Sociobiology: The New Synthesis*, Edward Wilson asserted that human sociobiology is 'nowadays also called evolutionary psychology' (Wilson, 2000, p. vii). However, Leda Cosmides and John Tooby, two of the founders of evolutionary psychology, have explicitly denied that their discipline draws greatly from Wilson's sociobiology, while others disagree. When two other leading evolutionary psychologists, Martin Daly and Margo Wilson, published an article in which they described evolutionary psychology as 'the work of all those engaged in evolutionary analyses of human behaviour' (Daly and Wilson, 1999), they incurred the wrath of colleagues Eric Alden Smith, Monique Borgerhoff Mulder, and Kim Hill, who did not identify with this point of view (Smith et al., 2000). Social scientists accused evolutionists of ignoring cultural explanations of human behaviour; yet advocates of the 'meme' perspective provided an evolutionary explanation that was exclusively cultural.

One of our goals with this text is to lead the reader through this minefield of terms and concepts. In truth, there are many ways of using evolutionary theory to study human behaviour, and there is much disagreement within the field as to the best way to do it. This diversity of approaches can result in confusion for outsiders, as well as for those who wish to use evolutionary theory themselves and are trying to distinguish between methodologies. What are the assumptions of each school of thought? Are some approaches more reliable than others? Are some right and others wrong? We discuss the history of using evolutionary approaches to describe human behaviour dating back to Darwin, which helps to explain why some of these divisions exist. Then, by comparing the different approaches, and critically evaluating their assumptions and methods, we hope to provide the information that the reader needs to assess which perspectives are the most compelling and what methods are the most useful.

Asking evolutionary questions

The Nobel Prize-winning ethologist Niko Tinbergen suggested that there are four principal types of question that can be asked about a behaviour pattern

(Tinbergen, 1963). Take an aspect of human maternal behaviour, for instance, breastfeeding. If one were investigating the behaviour of mothers towards their babies, a researcher could ask: (1) *'What hormonal mechanisms and cues from the infant elicit breastfeeding by the mother?'* (2) *'How does maternal care change over the lifetime of the mother as she becomes more experienced at raising children?'* (3) *'Does breastfeeding provide non-nutritional benefits, such as protection against disease and enhanced mother–infant bonding, that have led to it being favoured by natural selection, above and beyond nutritional value?'* (4) *'Why has the period of exclusive breastfeeding evolved to be shorter in human beings than in closely related primate species, and does the timing of weaning depend on whether fathers and other relatives regularly provide infant care?'* The first question explores the *mechanisms* or immediate causes underlying behaviour, whereas the second investigates the *development* of the behaviour during the lifetime of the individual. The third question addresses the *survival value*, or *function*, of the behaviour pattern and examines what advantage it gave our ancestors in the struggle to survive and reproduce. The fourth investigates the *evolutionary history* of the behaviour and asks how a particular species came to exhibit one trait rather than another. Questions of *function* and *evolutionary history* address different aspects of the evolution of a behaviour pattern.

In the book, we will see that different sub-fields place varying degrees of emphasis on the relative importance of these four classes of question. Disputes have arisen when protagonists have not clearly distinguished between these levels of analysis. As outlined by Tinbergen, we believe that answers are required on all of these dimensions to fully understand why a behaviour pattern occurs. One emphasis in the book will be that a full consideration of all four questions will provide the only complete description of human behaviour.

Another key issue to which we will return is the value of making comparisons across species. Knowledge of how other animals behave can be of value in interpreting human behaviour. For illustration, studies that have compared imitation in humans and chimpanzees have established that, unlike chimpanzees, humans copy actions that are not causally relevant to producing the desired outcome (Horner and Whiten, 2005). This between-species difference implies that there may be an additional social function to human imitation, or that our species is particularly tuned to detect and respond to the teaching intentions of others. However, we must bear in mind that behaviour patterns that at first sight appear to be similar in human beings and other animals may in fact be somewhat different. An example is the same-sex sexual behaviour observed in many animal species, which has frequently been described as 'homosexual' behaviour (Bagemihl, 1999). Although same-sex sexual behaviour has been extensively documented in animals, in most cases, its function and relationship to sexual orientation remains unclear (Bailey and Zuk, 2009). For example, same-sex mounting in male non-human primates appears to play a role in displays of dominance rather than providing a measure of sexual preference (Dixson, 2012). Thus, despite superficial similarities, the causes of these behaviour patterns may be quite different for humans and other species.

We can use another example to show what happens when evolutionary explanations are used to explain a trait before the relevant comparative evidence is well understood. In the 1970s, scientists started to ask, '*Why do women have concealed ovulation?*' Women exhibit no obvious sign that an egg has been released from an ovary and that they are approaching the time in the monthly cycle when sex is most likely to result in pregnancy. In contrast, female chimpanzees and baboons advertise their time of ovulation with bright red 'sexual swellings' around their genitalia (Street et al., 2016). When a female chimpanzee or baboon is fully swollen, males will compete for the chance to mate with her. Numerous evolutionary hypotheses were proposed to explain the evolution of 'concealed ovulation' in the human lineage. For example, males would be forced to 'guard' a female and, as a consequence, have more certainty of paternity and provide more help with offspring care. The problem with these hypotheses was that 'concealed ovulation' is probably not a derived trait in human beings. In other words, it is not 'concealed ovulation' among our ancestors, but 'advertised ovulation' in other species, that has evolved. Comparative analyses have shown that chimpanzees most likely evolved sexual swellings after splitting from their common ancestors with human beings (Nunn, 1999). Therefore, the wrong question was being addressed.

The study of human behaviour can derive much useful information from the behaviour of animals, particularly the other primates (e.g., Hrdy, 1999, 2009; de Waal, 2001). Indeed, a comparative analysis is a critical step towards determining which evolutionary question to ask. Experimental comparisons with other animals have shed light on many aspects of human cognition and behaviour, including memory, problem-solving, social cognition, and communication, sometimes pinpointing key differences and, at other times, establishing new similarities (Laland and Seed, 2021). Where a gulf was once perceived between 'rational man' and 'instinct-driven' animals, research has established some striking continuities. For instance, we now know that particular species of animals can make and use tools, manipulate symbols and plan for the future, and show many aspects of cognition once regarded as exclusively human (Kaufman et al., 2021). However, the examples of same-sex sexual behaviour and concealed ovulation discussed above also reveal how we must guard against superficial similarities between species and be wary of labelling a behaviour as an evolved trait without testing this assumption, as evolutionary analyses may sometimes be mistaken (Zuk, 2002).

Human culture, learning, and genetic determinism

The titles of popular science books taking an evolutionary perspective have described human beings variously as 'naked apes', 'lopsided apes', 'aquatic apes', or 'the third chimpanzee', and have referred to 'man the hunter' and 'mother nature'. Additionally, we have been told 'how the mind works', 'why sex is fun', and have had 'consciousness explained'. However, can there ever be a straightforward evolutionary explanation of human behaviour? Isn't there something different about human beings compared to our primate cousins and other animals? We have a complex culture, built around a spoken language and written texts. Can human behaviour be explained by biology alone? Doesn't our culture set us apart? For most social scientists, human behaviour is largely learned from other people. Consequently, the principal reason why, on average, people of New York differ in how they think and in what they do from Ache hunter–gatherers of Paraguay or Arctic Inuit of Canada is thought to be because they have been exposed to divergent cultures or had different social experiences. For social scientists, culture is most commonly regarded as a cohesive set of ideas, beliefs, and knowledge that exists in a completely different realm to biology. These researchers believe culture is the primary influence on human behaviour.

In contrast, many evolutionary-minded researchers think about culture more broadly as the product of an evolutionary process. In many animal species, individuals grow up in an environment that contains other individuals of the same species, and most primates exhibit complex societies. Moreover, many animals acquire skills and knowledge by learning from others, frequently adopting the 'cultural' traditions that characterize their population. A prominent scientific paper reported thirty-nine distinct behaviour patterns maintained as cultural traditions in some populations of chimpanzees but not in others, including distinct patterns of tool use, courtship behaviour, and even medicinal skills, with each population's cultural repertoire apparently handed down by one generation to the next (Whiten et al., 1999). Similar variation, if less extensive, has been observed in orangutans, capuchin monkeys, whales, dolphins, and several other species (Whiten et al., 2017). Of course, there are important differences between animal and human cultures, but there are likely to be some continuities between them too.

The five evolutionary approaches outlined in this book differ in the way in which they regard human culture and the importance that they attribute to it. We shall see that some regard human culture as shaped by genetic biases and predispositions, and these researchers stress that there is much more uniformity to human behaviour and society than is given credence by traditional social scientists. They argue that there are hidden commonalities that are found universally across all societies; for instance, all cultures are structured by statuses and roles, and possess a division of labour. Others think of culture as the outcome of an interplay between our unusually flexible developmental systems and particular aspects of the ecological and social environment, an interplay that typically results in adaptive human behaviour. Perhaps seemingly arbitrary traditions for hunting particular animals or food preparation habits are actually the optimal solution to these problems given local conditions. Still others conceive of culture as an evolutionary process in its own right, with human minds adopting variant ideas in a similar manner to how genes are selected in biological evolution. Maybe scientific theories or political ideologies change over time in an equivalent manner to biological evolution. Finally, we shall come across a group of biologists and anthropologists that, like most social scientists, see culture as socially transmitted information that passes between individuals but focus on the interaction between genetic and cultural processes. For instance, perhaps we are predisposed to learn to be right-handed, but the frequency of right-handedness varies across cultures because of society-wide differences in their tolerance of left-handers.

The alternative evolutionary approaches also express quite different conceptions of the relationship between genes, development, learning, and culture. Some researchers regard developmental processes, including our capacity to learn for ourselves, as tightly constrained by our genetic make-up. From this viewpoint, we are programmed to learn those behaviour patterns that enhanced our survival and reproduction in the evolutionary past, and society reflects these evolved imperatives. For instance, perhaps we are predisposed to acquire a fear of snakes or spiders because these creatures constituted very real dangers for our distant ancestors. Others regard development as much more flexible and learning as only loosely guided by our genes, so that these processes can generate behavioural outcomes that are unspecified by prior selection. For example, rather than evolving a specific dietary preference for fried fish or chocolate, maybe evolution has furnished us with a tendency to eat whatever happens to taste good, as our taste buds have evolved to detect foods with the energy and nutrients to promote health and well-being.

Differences in ideas about culture and learning will be highlighted in the later chapters.

One important point that needs to be made before we proceed is that using evolutionary theory is not the same as taking a genetic determinist viewpoint. 'Genetic determinism' is the belief that our genes contain blueprints for our behaviour that will always be followed and that constitute our destiny. Such a belief would run contrary to much that is known about how human behaviour develops. Where researchers talk about genetic influences on human behaviour, they do not mean that the behaviour is completely determined by genetic effects, that no other factors play a role in our development, or that a single gene is responsible for each behaviour. While much of evolutionary biology focuses on genetic inheritance, it does not follow that researchers believe that genes are the sole determinant of behaviour, and the vast majority take it for granted that multiple environmental influences will play a part throughout development. We will come across evolutionists who describe 'genes for' a particular trait, by which they mean genetic variation that, along with a multitude of environmental factors, affects a character. This shorthand has been criticized as misleading by other biologists, and it may sometimes lead researchers to underestimate the importance of developmental processes; an evolutionary perspective does not equate with a genetic determinist view of human behaviour.

After decades of debate about the relative importance of 'nature' versus 'nurture', researchers have, unsurprisingly, come to the conclusion that both nature (generally associated with genes) and nurture (typically representing environmental factors, learning, and culture) are important. So where do we go from here? Should biologists concentrate on determining how much of behaviour can be explained by genetic inheritance, while the social scientists are left alone to discuss human cultures and social structure? We think not. Most biologists have long rejected this dichotomous mode of reasoning (Ridley, 2003). Although we hear reports in the press that scientists have detected 'the gene for' some trait such as empathy or schizophrenia, this language is highly misleading. Genetic and environmental influences on human behaviour are like the raw ingredients in a cake mix, with development analogous to baking (Bateson and Martin, 1999). Just as nobody expects to find all the separate ingredients represented as discrete, identifiable components of the cake, nobody should expect to find a simple correspondence between a particular gene and particular aspects of an individual's behaviour or personality. Indeed, most researchers agree that the very idea that an individual's behaviour can be partitioned into nature and nurture components is nonsensical, as a multitude of interacting processes play a role in behavioural development

(Bateson, 2017; Fox Keller, 2010). From this perspective, a complete understanding of human behaviour will only result from studying human beings as animals developing in a rich social environment and immersed in complex cultural traditions.

Evolutionary perspectives on human behaviour

The history of using evolutionary ideas to interpret human behaviour is no dry and dreary chronicle of academic research. For a century and a half, evolutionary thinking has had a dramatic influence on how human beings regard themselves and on how societies structure their shared values, institutions, and laws, particularly within Western societies. In Chapter 2, we provide an overview of these historical events. We begin with Charles Darwin, who wrote at great length about human beings. Darwin accumulated vast evidence that the gulf in mental ability between human beings and other animals was not as great as hitherto believed, and he showed both that animals are capable of surprisingly intelligent behaviour and that humans exhibit hidden brutish tendencies. We will also meet one of Darwin's relatives, Francis Galton, a pioneering scientist who devised the methods for using identical twins to investigate genetic influences on human behaviour. However, Galton was strongly biased towards biological explanations for human behaviour and mental abilities, which provided the basis for his writings on eugenics and founded a social movement that years later was to result in discrimination and enforced sterilization. We shall see that Darwinian views on evolution were distorted into *social Darwinism* that applied a 'survival of the fittest' doctrine to social institutions and used erroneous evolutionary arguments to propose that socialism was harmful and to justify unrestrained capitalism.

We also see how evolutionary-minded anthropologists and biologists in the nineteenth century, confusing evolution with progress, made unfounded statements about the evolution of human societies and argued that some 'races' had reached a higher level of evolution than others. Darwinian ideas were also to have a major influence on the theories of human development within psychology. For instance, Sigmund Freud took Darwin's ideas of sexual selection and the 'instinct' to mate and used them to develop his concept of the libido, a core of chiefly sexual urges that Freud suggested are the major underlying force behind human behaviour. We then move into the twentieth century and discuss how evolutionary ideas influenced the conflict between ethologists and psychologists over the relative importance of 'instinct' and learning. In the 1960s, popular ethology books, such as Konrad Lorenz's *On*

Aggression and Desmond Morris's *The Naked Ape* were to introduce dubious and sensationalistic evolutionary arguments to the general public and create major furores. Although we also describe the many positive ramifications of evolutionary theories of humanity, this history helps us to understand why many people remain wary of applying evolutionary reasoning to humans and helps us to understand the backgrounds from which modern approaches emerged.

In Chapters 3–7, we present five more recent evolutionary approaches to the study of human behaviour. Rather than providing a comprehensive overview of each sub-field, we aim to give the reader a taste of each of the alternatives. In all cases, we explain how the sub-field arose and introduce the researchers who played important roles in its development. We then provide an account of the key ideas and methods that characterize the sub-field, and a description of some of the research carried out by practitioners that illustrates the reasoning, merits, and findings of that particular school of thought. Each chapter ends with a critical analysis of the assumptions and methods of the sub-field, in which we attempt an impartial evaluation of the arguments made and the tools used by those researchers, and we discuss the main criticisms that have been levelled against each approach.

Contemporary evolutionary perspectives on human behaviour began in the 1960s and 1970s with a series of exciting breakthroughs in the study of animal behaviour that precipitated a revolution in evolutionary thought. Important new ideas such as *kin selection, reciprocal altruism*, and *evolutionary game theory* emerged through the work of William Hamilton, Robert Trivers, and John Maynard Smith, and these ideas were to alter the course of zoology. In Chapter 3, we depict these novel theories and methods that came together under the term 'sociobiology' and were brought to the attention of many through the books of Edward O. Wilson and Richard Dawkins. We describe how these ideas were applied to human behaviour and evaluate the political and scientific outcry that ensued. The principal charge made by the critics of human sociobiology is also examined, namely, that researchers had devised simplistic and prejudicial hypotheses. We highlight important ideas that emerged from human sociobiology, such as the careful comparison of human behaviour with that of other animals, which can be seen to have enlightened our views on human behaviour to this day. Although human sociobiologists were accused of abusing science to reinforce traditional values, we will give examples of sociobiological research that challenged, rather than bolstered, stereotypes concerning human sex differences. Finally, we will describe how the field triggered the development of the four major contemporary approaches, namely human behavioural ecology, evolutionary psychology, cultural evolution, and gene–culture coevolution. Almost certainly because of the controversy that surrounded it, few of today's researchers describe themselves as 'sociobiologists'.

In Chapter 4, we describe the sub-field of human behavioural ecology that has continued to employ methods devised to study animal behaviour to ask questions about human beings. These investigators, many of whom have backgrounds in anthropology, are interested in exploring to what extent the differences in human behaviour can be explained as adaptive responses to the habitat in which they live. Human behavioural ecologists construct mathematical models to compute the optimal human behaviour in a given context on the assumption that this is what might have evolved. They then test the model's predictions, primarily by studying traditional societies such as hunter–gatherers and pastoralists. We will see that these researchers have tested the hypothesis that people choose food items in order to maximize their caloric returns and that they hunt in optimally sized groups. The researchers assert that they are able to predict whether parents will have another baby given knowledge of the number of children that parents already have and the parents' wealth. Human behavioural ecologists have even devised evolutionary explanations for why parents in modern, post-industrial societies may most effectively pass on their genes by having fewer children. But do people really behave in an adaptive or optimal manner? Early critics suggested not and alleged that the research programme of human behavioural ecology is fundamentally misguided because it investigates the current function of behaviour instead of testing hypotheses concerning the evolved mental processes that guide behaviour. We will investigate to what extent these concerns are warranted.

In Chapter 5, we introduce the sub-field of evolutionary psychology. These researchers are primarily academic psychologists interested in the evolved psychological mechanisms that underlie human behaviour and who see modern human beings as creatures adapted to the environments of our hominin ancestors. They use this idea to discuss how behaviour patterns that may have no apparent utility in our modern environment are more easily understood if we reconstruct how natural selection was acting in the past when our ancestors were hunter–gatherers. Evolutionary psychologists claim to have identified a number of mental adaptations that regulate human behaviour even in modern societies, such as a tendency to be particularly sensitive to individuals who might be cheating during social interactions or for men to be more violent than women. Researchers report that across all continents there are universal sex differences in the characteristics that men and women look for in a partner, with men seeking to mate with many young women, and women choosing to prioritize male partners that are wealthy and powerful. Evolutionary theory has been employed to provide explanations for such sex differences. However, this research programme has also attracted considerable criticism, as many observers fear that knowledge about our ancestors' way of life is insufficient to generate reliable hypotheses about the present.

In Chapter 6, we evaluate the sub-field of cultural evolution and investigate the hypothesis that culture exhibits its own evolutionary process. We describe how practitioners in the field of cultural evolution use both mathematical models and experimental techniques to investigate the rules underlying the propagation of acquired knowledge through imitation and other forms of social learning. The main idea here is that aspects of our behaviour and knowledge, such as particular skills, beliefs, or rituals, change frequency in a manner that can be understood and even predicted with knowledge of evolutionary principles. Selective processes are viewed as operating on cultural knowledge through a combination of behavioural innovation, which generates variation, and selective retention of the best ideas. We will present case studies demonstrating that cultural selection can hone the perfect arrowhead and may lead to the evolution of complex societies. Cultural evolution researchers maintain that, as the evolution of human languages throughout history sufficiently resembles biological evolution, similar theoretical methods can be applied to linguistic data to reconstruct the historical relations among languages and test hypotheses concerning past movements of peoples. Perhaps most surprising of all, some cultural phenomena are described as resulting entirely from random copying via a process that is the cultural equivalent of random genetic drift. However, critics have suggested that cultural change is too different from biological evolution to follow Darwinian rules, and we will consider the legitimacy of these concerns.

In Chapter 7, we introduce gene–culture coevolution, a quantitative science that shares many similarities with the field of cultural evolution. Consequently, they use mathematical models devised from population genetics to predict how cultural traits spread through human populations by social learning, and how genes and culture coevolve. For these researchers, recent human evolution is dominated by this coevolution of genes and culture, which generates new evolutionary mechanisms and transforms evolutionary rates. Their models examine how cultural practices can have implications for genetic evolution. For instance, while most Westerners can drink milk without getting sick, the majority of adult humans lack genetic variants that allow an enzyme to break down lactose in adulthood. Intriguingly, those adults that can consume dairy products typically belong to cultures with a long tradition of dairy farming, suggesting that the cultural practice of dairy farming created selection favouring individuals who could drink milk without becoming ill. Their models also provide new methods for partitioning the variance in human cognitive and behavioural traits. Genetic explanations for differences between people in characteristics such as intelligence are frequently proposed based on studies of identical and non-identical twins, but gene–culture analyses have challenged these findings. However, gene–culture coevolutionary methods are also subject to criticism. For instance, some social scientists have objected to the idea that culture can be analysed as if it is composed of discrete psychological or behavioural characteristics, while others have suggested that human culture has changed too quickly to have significantly shaped human minds and behaviour.

In Chapter 8, all of these sub-fields are brought together for comparison. Advocates of each approach often claim to have the foremost or the only valid perspective on evolution and human behaviour, and protagonists from different schools of thought sometimes scrap among themselves. However, which approach is best? Does each school exhibit strengths and weaknesses, or is one method superior to, or more legitimate than, the other? Could the different approaches be integrated into a single overarching perspective that synthesizes techniques from disparate schools, or are there fundamental incompatibilities such that if one school is right, another must be wrong? What exactly are the key differences of opinion, and how can they be resolved? After comparing the alternative views and examining their ideological and methodological differences, in the final chapter of this book, we assess to what extent it is possible to cross the boundaries between approaches and integrate them into a broad yet rigorous evolutionary science of human behaviour.

Sense and Nonsense endeavours to provide the reader with an informed account of alternative evolutionary perspectives in the hope that they will be better able to distinguish between them and to learn from them in a discerning manner. Having completed this book, we hope that readers will have acquired the necessary knowledge and skills to be able to evaluate evolutionary hypotheses concerning humanity for themselves and to make their own judgements as to what makes sense and what is nonsense.

2
A history of evolution and human behaviour

Few ideas have contributed as much to biological knowledge as Charles Darwin's theory of evolution by natural selection, and yet this revolution within biology is just the tip of the Darwinian iceberg. 'Natural selection' has proved an irresistible concept, with countless scientists, social scientists, politicians, and business leaders drawn to its explanatory power. Darwinian ideas have been a source of great inspiration in the arts and in literature, too (Donald and Munro, 2009; Ruse, 2017). Therefore, it is perhaps not surprising that, since publication of *On the Origin of Species by Means of Natural Selection, or the Preservation of Favoured Races in the Struggle for Life* (Darwin (1859), there has been a long history of using evolution to interpret human behaviour and society, some of which makes distinctly disturbing reading. As John Maynard Smith (1975) pointed out, 'Attempts to import biological theories into sociology, from social Darwinism of the 19th century to the race theories of the 20th, have a justifiably bad reputation.'

In this chapter, we trace the history of applying evolutionary approaches to human behaviour in Western science and society from the 1850s to the 1960s. We will see that evolutionary ideas were important in shaping our concept of human behaviour, sometimes bolstering racism and sexism, while at other times dispelling unjust views. In fact, the last century and a half has been characterized by constant battles between evolutionary advocates and their critics. These battles have frequently coincided with a regular swinging of the pendulum to favour explanations for human behaviour that emphasize nature or nurture. We illustrate how evolutionary arguments have been put forward as pretexts to justify the eugenics movement, Nazism, unfettered capitalism, racist immigration policy, and enforced sterilization, as well as to argue that some 'races' were more advanced than others. The vast majority of these assertions employed crude distortions of Darwin's theory. As we will see, these distortions derive more from the work of other nineteenth-century intellectuals such as Jean-Baptiste de Lamarck and Herbert Spencer, although it is Darwin's name that is often unfairly linked to these views. We will also describe efforts by academics to counter racism and prejudice in society, and we reveal countless important scientific insights and advances that followed from an evolutionary viewpoint. However, the abuses of evolutionary theory are more often remembered.

This historical perspective provides a context within which we can begin to interpret contemporary disputes over the use of evolution. For instance, it helps us to understand why many social scientists are so resistant to evolutionary hypotheses. It also helps to explain why Edward Wilson's *Sociobiology* (1975a) was to provoke such profound hostility as to culminate in water being poured over him by protestors (see Chapter 3) and perhaps why many contemporary evolutionary psychologists place emphasis on the universal features of human behaviour (see Chapter 5). In the remaining part of the book, we will show how modern evolutionary approaches have increased our understanding of human behaviour, although we should remain aware of the social impacts that scientific theories can have.

Darwin's views on human behaviour

Within Western science, the history of using evolutionary theory to make sense of human behaviour began in earnest with Charles Darwin. Although questions about the origin and transformation of species were being debated for many centuries before Darwin's birth (Stott, 2012), Darwin was the first person to come up with a credible explanation for evolution, namely the process of *natural selection*. In *The Origin of Species*, Darwin (1859) patiently explained in a series of logical steps how natural selection works. Struck by the view of Thomas Malthus—that population growth would eventually reach a point at which insufficient food was available—Darwin suggested that those individuals in the population whose anatomical, physiological, and behavioural characteristics best fitted the environment would have the greatest chances of surviving and reproducing. If those characteristics were heritable, then the next generation would contain a higher frequency of individuals with these 'fitter' traits, and hence the population would change over time. At the time of publication, the dominant view of the natural world was that each species had been individually created and was immutable. The ability of natural selection to explain how variation among individuals may lead to adaptation of species to their environments and the origin of new species has been confirmed by countless studies and is now beyond dispute.

The striking feature of *The Origin of Species* is that Darwin does not mention human evolution, except to say the following in the final pages:

> In the distant future I see open fields for far more important researches. Psychology will be based on a new foundation, that of the necessary acquirement of each mental power and capacity by gradation. Light will be thrown on the origin of man and his history. (1859, p. 458)

While an eager public had to wait over a decade for Darwin to elaborate on this enigmatic statement, the idea that human beings had evolved became a source of intense public interest and hostility in the intervening years. However, Darwin, fearing persecution and ridicule, refused to be drawn further on human origins until a watertight case could be made. Instead, Darwin's great supporter Thomas Huxley tenaciously fought his corner for him. Although historians now suspect that the story that Huxley 'trounced' Bishop Wilberforce in a famous debate at Oxford University in 1860 has been greatly embellished (Brooke, 2001), the fact that eminent scientists like Huxley stood up to authorities, such as the Church, was undoubtedly critical to the spread of Darwinism. Huxley presented numerous lectures and, in his book *Evidence as to Man's Place in Nature* (1863), he used the skeletons of apes to provide undeniable evidence that human beings were of animal ancestry. Among archaeologists, the hunt for the fossilized remains of the ancestral 'missing link' between humans and other apes had truly begun.

By 1870, Thomas Huxley, referred to as 'Darwin's bulldog', had become 'the prophet of the new world of science' (Desmond, 1997). Indeed, largely through Huxley's efforts, being a 'scientist' became a legitimate profession, and 'science' came to exert a major political influence, with Darwinism providing the focus of this development. By the 1870s, Darwin was famous, and everyone was waiting to hear what the great man had to say on human evolution. With characteristic caution, Darwin eventually published *The Descent of Man, and Selection in Relation to Sex* (1871) and *The Expression of the Emotions in Man and Animals* (1872). Rather than dwelling on human anatomy, which had been Huxley's domain, Darwin drew attention to the question of the evolution of mental ability, for which there seemed to be a much greater divide between human beings and other animals. He maintained that there was variation in mental capacity both within and between species, and he suggested that being intellectually well endowed was advantageous in the struggle to survive and reproduce. In *The Descent of Man*, Darwin stated:

> To avoid enemies, or to attack them with success, to capture wild animals, and to invent and fashion weapons, requires the aid of the higher mental faculties, namely, observation, reason, invention, or imagination. (1871, p. 327)

Darwin sought to demonstrate that the differences in mental ability between human beings and other animals were not as great as generally believed. Darwin attempted to counter the widespread belief that animals were merely machines driven by in-built mechanisms, while human beings alone were capable of reason and advanced mental processing. He attacked this dichotomy from both sides, arguing that human beings had more brutish tendencies, and animals more elevated intelligence, than hitherto conceived. In contrast, Alfred Russel Wallace (1869), who had also struck upon the idea of natural selection after reading the works of Malthus, concluded that the complex language, music, art, and morals of human beings could not be explained solely by natural selection and must have resulted from the intervention of a divine creator during human evolution (Raby, 2001).

In the first part of *The Descent of Man*, Darwin documented the evidence that human beings have a number of behavioural characteristics in common with other animals, including 'self-preservation, sexual love, the love of the mother for her new-born offspring, and the power possessed by the latter for suckling' (1871, p. 36). Similarly, in *The Expression of the Emotions* (1872), Darwin catalogued an array of equivalent facial expressions in humans and other animals. By pointing out the striking similarities between the expressions associated with particular emotions across species, he dismissed the theory that expressions had been uniquely given to humans to communicate their emotional states to others. For instance, Darwin noted that apes and monkeys, similar to humans, have 'an instinctive dread of serpents' and will respond to snakes with the same screams and the same fearful faces as many of us do. He described how one day he placed a stuffed snake into the monkey enclosures at London Zoo and the poor animals 'dashed about their cages and uttered sharp signal-cries of danger, which were understood by the other monkeys' (1871, p. 43). Darwin also noted, around a century before modern researchers (Goodall, 1986), that chimpanzees use stone tools to crack open nuts, which suggested even less of a gap between the mental lives of human beings and apes than many in Victorian Britain wished to believe.

Darwin described the emotional lives of other animals in distinctly human terms. Even invertebrates were believed to feel pleasure and pain, happiness and misery, and exhibit some intelligence. He maintained that, for all animals, 'terror acts in the same manner on them as on us, causing the muscles to tremble, the heart to palpitate, the sphincters to be relaxed, and the hair to stand on end' (1871, p. 39). He suggested that, similar to puppies, young ants will chase and pretend to bite each other, and he argued that they are excited by the same emotions as we are. He also maintained that courage and timidity are seen in dogs, and that horses can be sulky, and monkeys vengeful. Judged by contemporary standards, these arguments are naïve, anthropomorphic, and anecdotal. Yet many of Darwin's most fundamental assertions about the mental abilities of animals have been proven to be correct. Most animal behaviour researchers agree that many animals do feel pleasure and pain, that they are capable of learning and intelligent behaviour, and that they probably do share many of the same emotional states as human beings (Rose, 2020). Darwin adopted an anthropomorphic style with the intention of showing that emotions and expressions were not unique to human beings, while his comparisons of the expressions of emotions in different human populations supported his view that such expressions are shared universally across cultures.

The publication of Darwin's three great works had a lasting impact on the emerging field of Western psychology. At the time, this field was dominated by physiologists who were investigating the mechanisms of the brain, and by philosophers who were creating theories about how the mind works. From the seventeenth to the early part of the nineteenth century, British philosophers such as John Locke, David Hume, and John Stuart Mill had argued that the human mind at birth is like an empty box, or blank slate (*tabula rasa*), that is free of in-built knowledge and is gradually filled as we experience the world. Eventually, our ideas and observations become integrated so that we can make sense of that around us, an idea that became known as *associationism*. With hindsight, we can see how this theory must be incomplete. Human minds cannot construct a mental picture of the world unless they have ready-built structures that make knowledge acquisition possible. In his famous *Critique of Pure Reason* (1781), the German philosopher Immanuel Kant made the point that there must be certain preconditions to the human mind that contribute to our conception of the world, and Kant's insights have been confirmed by a vast array of recent findings from psychology, neuroscience, and artificial intelligence. We now understand that the mental apparatus that allows us to perceive, interpret, and model the world around us is partly a product of our genes. Darwin's writings were partly instrumental in bringing about a decline in associationist views within psychology.

In the second part of *The Descent of Man*, Darwin introduced the concept of *sexual selection* to provide an additional explanation for physical and mental differences between the sexes. Following the principles of natural selection, this idea stated that some characteristics may have evolved that increase an individual's chances of gaining matings, either by enhancing competitive abilities among members of the same sex (usually presumed to be more important in males than in females) or by enhancing the likelihood of being chosen as a mate (usually viewed as females choosing particular males). Such factors were suggested to generate selection for traits and characteristics in one or the other sex, such as the peacock's extravagant tail or the large antlers of male deer. Human beings were firmly included in Darwin's theorizing about sexual selection from the beginning, and the theory was developed not just to explain the differences between men and women but also the physical and behavioural variation between human populations that Darwin observed on his *Beagle* voyages (Desmond and Moore, 2009).

Today, Darwin's views on mental differences between the sexes in human beings certainly appear outdated. He wrote that, 'Man is more courageous, pugnacious, and energetic than woman, and has a more inventive genius' (1871, p. 316) and that, 'male monkeys, like men, are bolder and fiercer than the females' (1871, p. 320). Based on these assumptions, Darwin suggested:

> These characters will have been preserved or even augmented ... by the strongest and boldest men having succeeded best in the general struggle for life, as well as in securing wives, and thus having left a large number of offspring. (1871, p. 325)

However, on the downside, Darwin suggested that, 'Man delights in competition, and this leads to ambition which passes too easily into selfishness' (1871, p. 326). Furthermore, Darwin suggested:

> Woman seems to differ from man in mental disposition, chiefly in her greater tenderness and less selfishness. ... It is generally admitted that with woman the powers of intuition, of rapid perception, and perhaps of imitation, are more strongly marked than in man. (1871, p. 326)

Darwin may be forgiven to some extent if he is compared with the prevailing views of Victorian Britain. His observation that, with education, 'woman should reach the same [intellectual] standard as man' (1871, p. 329) suggests that his views were more liberal than those of many others during that period. Also, Darwin's proposal that female choice plays a crucial role in the process of sexual selection also acknowledged a more active role for women in their sexual relationships and sexual behaviour than was commonly permitted at the time.

Similarly, although Darwin's ideas on racial differences among human populations published in *The Descent of Man* are prejudiced by today's standards (Fuentes, 2021), he was again willing to consider that opportunity and experience plays a major role in such differences. On his voyage on the *Beagle* from 1831 to 1836, Darwin travelled with three Fuegians from South America (Palma, 2023). On a previous voyage, two of them had been taken hostage in reprisal for a stolen boat, and the third had been 'bought', or forcefully taken, in exchange for a pearl button, after which he was named. The three had been transported to England, 'educated' into British society and Christianity for a year, and they were now being returned to their homeland as intended emissaries. Jemmy Button (whose real name was Orundellico) made a particularly strong impression on Darwin and fellow shipmates, reportedly because of his linguistic abilities, manners, and good humour. Darwin was stunned when he arrived in Tierra del Fuego, as the other natives appeared to him 'wretched' and 'wild' in comparison. This personal experience brought it home to Darwin that many differences between peoples were brought about by culture and the environment and that a change in circumstance could have far-reaching effects. As we shall see, many of Darwin's contemporaries maintained a different attitude, assuming that purported sex and race differences in mental abilities were inevitable and could never be altered by changed opportunities.

Subsequent careful scrutiny of the private notebooks of Darwin has revealed that many of the ideas in *The Descent of Man* were conceived as far back as 1838. His notebooks touched on a wide range of psychological topics, including memory, learning, imagination, language, emotion, and psychopathology. In his notebook of 16 August 1838, Darwin proclaimed that 'he who understands baboon would do more towards metaphysics than Locke'. By this, Darwin meant that the study of animal behaviour would be more useful than philosophy in helping us to understand the human mind. This rather startling claim sounds extraordinarily similar to some of the bold statements that were to emerge from the field of human sociobiology a century and a half later (see Chapter 3), but, in Darwin's case, his published work was far more considered and judicious than his private writings. Nonetheless, Darwin's view that psychology and the study of human behaviour should be based on an understanding of biology and the concepts of variation, heredity, and adaptation had a far-reaching impact.

Galton and the development of eugenics

Francis Galton was Darwin's younger half-cousin (they shared the same grandfather, Erasmus Darwin) and was one of an inner circle of British intellectuals who were privy to his thoughts prior to *The Origin*'s publication. The emphasis that Darwin placed on heredity and individual differences provided a source of inspiration for Galton's work, which sought to explain why people differ in mental ability. His major work, *Hereditary Genius*, was published in 1869. In this book, Galton traced the genealogies of 'able families' among the judges of England, the peerage, military commanders, and men of science, literature, poetry, and music. For instance, Galton noted that the Bachs were all tremendous musicians, while Darwin's family (from which he modestly excluded himself) were great scientists. Using this information, Galton suggested that mental abilities were inherited, as opposed to the prevailing view at the time that the human mind acted 'independently of natural laws'.

Galton was a polymath who made major contributions to mathematics, psychology, and evolutionary theory, and who pioneered the use of identical twins in the study of genetic influences on behaviour (Bulmer, 2003). One of his lesser-known accomplishments was the invention of fingerprinting to help the police solve crimes. He also set up an anthropometric laboratory to undertake the collection of physical and mental testing of men and women. Galton became obsessed with measurement; for example, as he travelled around Britain, he secretly constructed a 'beauty map' of the cities, concluding that the incidence of attractive women was highest in London and lowest in Aberdeen. He also enlivened dull scientific meetings by attempting to measure the boredom level of the audience, eventually settling on a measure of fidgets per minute, a study that he published in the eminent journal *Nature*.

However, Galton also exhibited extraordinary prejudices. He believed, for example, that some men belonged to the criminal type and that no amount of environmental improvement would alter this fact, and that the inability of women to distinguish the merits of various wines confirmed the inferiority of female intellectual ability. He tended to ascribe almost all differences between human beings to heredity, what we would now call genes, and virtually nothing to education or opportunity. His hereditarian bias is manifest in his definition of genius, which was 'an ability that was exceptionally high *and at the same time inborn*' (Galton, 1869, italics added). He acknowledged that education could develop the mind's full potential; however, individuals could never rise above their inherited mental capacity. Hence, Galton was opposed to the education and suffrage of women. Galton also exhibited racial prejudices; for example, he regarded Africans as having a lower average mental ability than Europeans. In the chapter 'The Comparative Worth of Different Races', Galton suggested, 'Every long-established race has necessarily its peculiar fitness for the conditions under which it has lived, owing to the sure operations of Darwin's law of natural selection' (1869, p. 336). Galton maintained that, with time, the 'more civilized' races would inevitably eliminate native populations because of the latter's inability to cope mentally with the tasks of the supposedly superior society.

Galton thus became increasingly concerned for the future intellectual quality of humanity, for instance fearing that, in Britain, the lower classes would outbreed the gentry. In an article written in 1894, he stated that 'It has now become a serious necessity to better the breed of the human race. The average citizen is too base for the everyday work of modern civilisation' (cited in Forrest, 1974). In *Hereditary Genius*, Galton stated, 'It seems to me most essential to the well-being of future generations, that the average standard of ability of the present time should be raised' (1869, p. 344). To accomplish this goal, he suggested the active encouragement of early and judicious marriage by those possessing 'favourable hereditary qualities', and for the weak and criminal to be sent to celibate monasteries. Thus arose Galton's theory of *eugenics*, later defined by him as 'the science which deals with all influences that improve the inborn qualities of a race' (1904, p. 1). Although Darwin reported the work of Galton in *The Descent of Man*, he did not condone these views, instead stating that the intentional neglect of the weak and helpless would be a 'certain and great present evil' (1871, p. 169).

For Galton, eugenics became a great passion. Ironically, his book *Inquiries into Human Faculty and Its Development* (1883), which set out his eugenic ideas, was criticized at the time of publication mainly for its anti-religious views rather than the eugenics content. Using his information on family histories, Galton stunningly concluded the inefficacy of prayer, by showing that men who were much prayed for, such as those high up in the church, did not live longer than those at the top of other professions, such as law, and that ships bearing missionaries sank just as often as those carrying material goods. Galton was now regarded as one of the world's leading psychologists, and his personal views were having widespread influences beyond academic circles (Bulmer, 2003). For example, within the first few decades of the twentieth century, vibrant eugenic movements had developed in more than thirty countries, including Britain, the United States, and Germany, and, by 1914, laws preventing the marriage of the 'mentally deficient' on eugenic grounds had been introduced in thirty American states. Eugenics became a mainstream view within both evolutionary science and society for much of the first half of the twentieth century, and its insidious effects can be felt even through to the present day (Sussman, 2014).

The ascent of progressive evolution

Towards the end of the nineteenth century, Darwin's theory of natural selection was losing favour as an explanation for evolutionary change. The theory was partly hindered by the lack of knowledge of genetics. However, the main opposition to natural selection came from physicists such as Lord Kelvin. His calculations seemed to show that the Earth was not old enough to have supported life for the thousands of millions of years that would have been required for species to evolve via natural selection. While Kelvin was convinced that the Earth was no more than 100 million years old at most, this estimate is now known to have been incorrect, as the Earth is approximately 4.5 billion years old; however, by 1870, the evidence appeared stacked against natural selection. In comparison, the earlier teachings of another great evolutionary thinker, Jean-Baptiste de Lamarck, were in the ascendancy. Lamarck published his major work on evolution, *Philosophie Zoologique* (1809), during his tenure as a professor at the National Museum of Natural History in Paris, where he also gave an annual series of public lectures. Lamarck suggested that all species were independently created and could be placed on a scale with the most similar species next to each other. Each species could then move up Aristotle's 'chain of being' that culminated in human beings. The process by which this progression was thought to occur was the inheritance in offspring of characteristics acquired by parents during their lifetime, such as the passing on of stretched joints or well-exercised muscles. Lamarck's view of evolution was linear and progressive, with species exhibiting an inherent striving to evolve ever-greater complexity, with the pinnacle of life on Earth being human beings.

The theory that species have transformed over time from more humble ancestors was initially rejected in France, largely owing to the opposition of the powerful comparative anatomist Georges Cuvier, whereas in Britain it was regarded as dangerously atheistic and too closely linked with French revolutionary ideas. Indeed, Darwin's emphasis on gradualism was probably partly an attempt to dissociate evolution from revolution. Lamarck died in poverty and, while his zoological writings on invertebrates remained well respected, his work on evolution fell into scientific disrepute. At his funeral, his daughter is said to have cried out, 'My father, time will avenge your memory!' (Boakes, 1984). She was partly right in two regards. Firstly, throughout Europe and North America, Lamarck's view that species change through time was gaining a growing number of supporters and now underpins modern evolutionary theory (Björklund, 2019; Stott, 2012). Secondly, when the nineteenth-century physicists stated that there was insufficient time for natural selection to do its work, the 'inheritance of acquired characters' seemed to fit by providing a fast evolutionary process. While the inheritance of acquired characteristics has been widely discredited in the twentieth century, related ideas, such as epigenetic and cultural inheritance, are now receiving considerable attention (see Chapter 8).

One advocate of Lamarckian ideas, Herbert Spencer, was particularly influential in the mid- to late nineteenth century (Oldroyd, 1983). Spencer was born in England in 1820 and, although he initially trained as a civil engineer, his major interests were sociology, psychology, and philosophy. Spencer cultivated the idea that all things change inevitably from a simple to a more complex state, including species and human societies. For instance, in an article titled 'Progress: its law and cause', Spencer wrote:

> The advance from the simple to the complex, through a process of successive differentiations, is seen alike in ... the geologic and climatic evolution of the Earth, and of every single organism on its surface; it is seen in the evolution of Humanity ...; ... in the evolution of Society in respect alike of its political, its religious, and its economical organization; and ... in the evolution of all those endless concrete and abstract products of human activity which constitute the environment of our daily life. (1857, p. 465)

In the modern era, we might be tempted to say that characterizing evolution as a progression from simple to complex is a distortion of Darwinian ideas. However, in fact, the term 'evolution' long predates Darwin's time, and the progressive connotation is actually closer to the original meaning; 'evolution' derives from the Latin word 'evolutio', which means unrolling. In ancient Rome, books were written on lengths of parchment and rolled onto wooden rods, so they had to be unrolled to be read (Caneiro, 2003). From the seventeenth century, 'evolution' was widely deployed to refer to an orderly sequence of events, particularly one in which the outcome was somehow contained within it from the start. Spencer's concept of mental evolution was that of a single continuum from the reflexes of simple animals to their pinnacle in the intelligence of the civilized man. In his influential *Principles of Psychology* (1855, 1870), which was first published prior to Darwin's *Origin of Species* (1859), Spencer described how human societies gradually became more developed, with the 'large-brained European' depicted as being mentally far in advance of 'primitives'. A similar view, that human societies progress through various levels punctuated occasionally by revolutions that take a society to a higher level, was being propounded by Karl Marx.

In the late nineteenth century in America, Spencer's views of evolution and society rivalled Darwin's for popularity and were endorsed by religious and business leaders (Oldroyd, 1983). When Spencer travelled to America in 1882, he was warmly greeted and his books were bought by the thousands, as his views appeared to justify the business ideas of the newly wealthy country. Although Spencer himself predicted that societies would progress to a stable, benevolent state (Francis, 2007), his slogan 'survival of the fittest' was eagerly adopted by business, where it was assumed that competition was inevitable and fitness should be measured in terms of wealth. The extent to which evolutionary thinking was being applied to all aspects of human life is exemplified by Henry Drummond's (1894) book, *The Ascent of Man*, which provides evolutionary accounts of human morality, language, cooperation, family structure, and weapon production. The endorsement of evolutionary ideas by society and business began the movement known as 'social Darwinism', although 'social Spencerism' would be a more appropriate term for accounts that rely on the notion of progress, as these ideas derived far more from Spencer.

Darwin (1859) portrayed evolution more like a branching tree than a ladder. In contrast, because social Darwinists erroneously believed that biological evolution was progressive, they concluded that competition should be actively encouraged. Social Darwinism was used to justify doctrines such as social conservatism, militarism, eugenics, laissez-faire economics, and unfettered capitalism (Oldroyd, 1983). The leading proponent among American academic circles was William Sumner, Professor of Political Economy at Yale. Sumner asserted, 'Millionaires are a product of natural selection ... They get high wages and live in luxury, but the bargain is a good one for society' (cited in Oldroyd, 1983). In contrast, socialist schemes were regarded as a menace to society as they 'promote the survival of the unfittest'. Business leaders, such as Andrew Carnegie and J. D. Rockefeller, also exploited evolution to their own ends. For example, Carnegie argued that 'the concentration of business in the hands of the few ... was essential to the future progress of the race' (cited in Oldroyd, 1983). This reasoning is a gross distortion of Darwinian thought, and Darwin had wholly rejected such interpretations of his ideas.

During the last two decades of the nineteenth century, social Darwinism thrived in Europe and North America, partly because the idea that nature counted much more than nurture in the expression of human behaviour appeared to justify widespread structural inequalities. The huge contrast between the power and wealth of Western nations compared to the rest of the world came to be seen as reflecting in-built differences in the psychology and abilities of different 'races'. This view also gained support from prominent scientists. For example, Ernst Haeckel, an eminent German professor of zoology, championed the view that the evolution of a species progressed through increasingly higher or more advanced stages, similar to the stages that are observed during an individual's development. Inspired by his studies of vestigial traits that emerge and disappear during specific stages of embryonic development, Haeckel (1866) canonized this idea with his 'biogenetic law', which suggested that ontogeny (development from conception to death) is a re-enactment of phylogeny (the evolutionary history of the species). Haeckel had been converted to evolutionary thinking on reading *The Origin of Species* and was an energetic recruit, writing a series of papers and books on evolution that established him as one of the world's leading evolutionists. However, like Spencer, Haeckel's view of evolution tended more towards Lamarck's, and Haeckle's law proved to be an oversimplification of how embryos develop.

For Haeckel, evolutionary theory also had very definite political implications and provided the framework for his commitment to the reform of political institutions and to the unification of Germany. He used his immense authority in German-speaking countries to promote views on inherent racial differences right up to the First World War. Historians have noted a strong lineage of ideas passing directly from Haeckel to the appalling doctrines of Nazi theorists (Oldroyd, 1983). Years later, pseudoevolutionary political diatribe was to reach its diabolical zenith with the publication of *Mein Kampf*, in which Adolf Hitler drew on an erroneous conception of 'blending inheritance', on facile analogies from animals, and on Spencer's 'survival of the fittest' doctrine to give a quasi-scientific argument for the need for racial purity. In biological terms, Hitler's arguments were nonsensical, yet no body of work illustrates more dreadfully how dangerous is the distorted view of biological evolution as progress.

Even George Romanes, chosen by Darwin as his intellectual successor (Boakes, 1984), regarded evolution in progressive terms. A strong friendship had developed between the two men after Romanes began writing to Darwin in 1874. Romanes addressed the question of how human mental abilities had evolved by comparing human and animal behaviour. The animal mind was a popular topic in Britain in the 1870s, and countless letters flowed into scientific and popular journals reporting striking observations on the mental capabilities of animals. Romanes collected these anecdotes and summarized them in his 1882 book *Animal Intelligence*. This treatise collates countless astounding examples of animal behaviour arranged in order of mental ability, from the earwig that had been trained to climb up the curtain every day to eat breakfast to the dog that understood the mechanical principle of the screw. However, Romanes appears to have been more influenced by Spencer and Haeckel than by Darwin, citing both frequently. Using Haeckel's biogenetic law, Romanes placed the mental abilities of animals on an ascending scale, culminating in humans. He then proposed that, during development, a human being plays out this evolutionary ladder. Each age was ascribed a comparable animal intelligence (Table 2.1); for example, at 3 weeks of age, a baby was roughly equivalent to an insect in mental ability; at 4 months of age, an infant was equal to a reptile; and by 15 months, the youngster was usually brighter than apes and dogs.

The idea that human societies progressed through various levels was also prevalent within the emerging field of anthropology (Harris, 2001; Oldroyd, 1983). Even the British anthropologists John Lubbock and Edward Tylor, with whom Darwin was most closely associated, had no doubt that higher cultures were associated with more advanced races whose members had larger and more effective brains. Lubbock and Tylor argued that all civilized nations are the descendants of barbarians, first, because some traces still existed in customs and language, and also in archaeological remains such as flint tools, and second, because 'savages' were sometimes independently able to raise themselves in the scale of civilization. Tylor set out this theory in his two major publications, *Researches into the Early History of Mankind and the Development of Civilization* (1865) and *Primitive Culture* (1871). He reasoned that, if one studied small-scale societies in other parts of the world, one could gain historical insights into the past Stone Age cultures of Europe. In 1877, in his book *Ancient Society, or Researches in the Lines of Human Progress from Savagery*

Table 2.1 George Romanes' (1882) depiction of the relative intelligence of humans at various stages of mental development and the corresponding level of other species

Human development	Equivalent to	Psychological ability
Sperm/egg	Protoplasmic organisms	Movement
Embryo	Coelenterata	Nervous system
Birth	–	Pleasure and pain
1 week	Echinodermata	Memory
3 weeks	Larvae of insects	Basic 'instincts'
10 weeks	Insects and spiders	Complex 'instincts'
12 weeks	Fish	Associative learning
4 months	Reptiles	Recognition of individuals
5 months	Hymenoptera	Communication of ideas
8 months	Birds	Simple language
10 months	Mammals	Understanding of mechanisms
12 months	Monkeys and elephants	Use of tools
15 months	Apes and dogs	Morality

Table 2.2 Lewis Henry Morgan's (1877) stages of cultural evolution

Stage	Characteristics
Lower savagery	Fruit and nut subsistence
Middle savagery	Fish subsistence and use of fire
Upper savagery	Bow and arrows used as weapons
Lower barbarism	Pottery used
Middle barbarism	Animals domesticated, maize cultivated with irrigation, adobe and stone architecture
Upper barbarism	Iron tools used
Civilization	Phonetic alphabet and writing employed

through Barbarism to Civilization, the American anthropologist Lewis Henry Morgan took this viewpoint to its logical end point by documenting the stages of cultural evolution through which societies were assumed to progress (Table 2.2).

These anthropologists argued that all human races shared a common ancestor, but that some races were higher on the scale of progression than others. This view was in contrast to the extreme ideas of another set of anthropologists, who argued that slavery was 'natural' because different races were actually different species. This latter group, typified by the phrenologist Samuel George Morton and backed by the influential Harvard zoologist Louis Agassiz, spent their time in the physical description and classification of different races from around the world (Desmond and Moore, 2009). Such racism relied on the idea that human beings within a population could all be described as a particular 'type'. Yet Darwin's view of evolution crucially highlighted the importance of variation within populations and rejected such typological thinking (Mayr, 1982). The evolutionary evidence clearly supported the single-species view. Thomas Huxley argued that the ability of all humans to interbreed demonstrated we must be one species, and Darwin's work on the similarities between races in mental abilities and expression of the emotions clearly supported this view.

The widespread idea that British and North American societies were superior to those of other cultures provided even greater impetus to the social Darwinist movement (Boakes, 1984). For instance, social institutions in Victorian Britain were presumed to be natural, good, and healthy, whereas 'primitive' societies were caricatured as being abnormal and degenerate. During the 1890s, a number of prominent biologists, including Huxley in Britain and James Mark Baldwin in the United States, reacted angrily to what they saw as the damaging use of evolutionary theory to justify obnoxious social and ethical values. Unfortunately, their protestations fell mainly on deaf ears. Other voices from outside of the predominantly White spheres of European and North American scholarly societies also had their voices ignored on the topic of racial equality. For instance, in his book, *De l'Égalité des Races Humaines* (1885), Anténor Firmin critiqued the work of physical anthropologists who were assuming the superiority of particular races over others (Fluehr-Lobban, 2000). Firmin was a Haitian anthropologist who interacted with the anthropological society of Paris while based there as a diplomat, and his forward-thinking book outlined a new, comprehensive science of anthropology that influenced many African scholars. Yet, his work remained relatively unknown within Western academic literatures for over a century. As with other non-White, and female, scholars or would-be scholars, their contributions and viewpoints remained largely unheard. Significant developments in the history of anthropology would have to wait until the early part of the twentieth century, when debates about the relative roles of nature and nurture had been played out in the fields of zoology and psychology.

The nature–nurture debate

The leaders of Victorian society in Britain kept a close and informed interest in current scientific developments, including zoology. Consequently, public attention was drawn to a set of extraordinary experiments carried out on birds in the 1870s by a young British scientist called Douglas Spalding (Boakes, 1984). Originally earning a living mending slate roofs, Spalding educated himself by attending lectures in Aberdeen and London. He became frustrated that the leading psychologists and philosophers were prepared to discuss whether the mind was, or was not, influenced by 'instincts' without ever testing these assertions. Spalding began to carry out his own set of experiments on chickens to investigate whether any inherent abilities were present at hatching. To test whether a young chick was able to move about its world without bumping into objects, peck accurately, and locate sounds without any prior sensory experience, he removed sections of shell from eggs just before the birds emerged, put wax in their ears, and covered their eyes with a patch to remove any auditory or visual cues, and then tested them a few days after hatching when the wax and hoods were removed. He inferred that these birds were just as capable as other birds of pecking accurately, making coordinated movements, avoiding objects, and responding appropriately to threats like a hawk and concluded these abilities must be 'instinctive'. These experimental studies raised the idea that human behaviour might similarly depend upon instinct.

Spalding also discovered that chicks would *imprint* (i.e. form an attachment) on the first object that they saw after hatching, which is usually their mother, and would 'instinctively' follow her around. The ability of chicks to imprint was found to be greatest during the first few days of life, which meant that imprinting had occurred during an early, 'sensitive' period of development. Spalding conducted his studies when he was employed by Lord Amberley, the son of the British Prime Minister, as a tutor to his eldest son, and he was encouraged to continue his pioneering research in their house with Lady Amberley as his assistant. Unfortunately, Spalding's research ended suddenly. After the deaths of Lord and Lady Amberley, and to the consternation of their powerful grandfather, the guardianship of the young sons was left to Spalding, which led to questions being asked about the paternity of the children. The guardianship was fiercely contested, and Spalding was forced to emigrate to France, where he died a year later.

Back in Britain, emphasis continued to be placed on the comparison between human mental abilities and those of other animals. The psychologist Conwy Lloyd Morgan, who studied under Thomas Huxley, opposed Romanes' anecdotal approach to the study of the mind and undertook his own comparative study of instinct. As one of the founding fathers of both ethology and comparative psychology, Lloyd Morgan wrote fourteen substantial books, including *Habit and Instinct* (1896), *Animal Behaviour* (1900), and *The Animal Mind* (1930). In these books, he argued that instinctive behaviour could be identified through detailed observational studies of animal movement patterns, and that such behaviour patterns could be modified by experience. Lloyd Morgan also propounded the notion of using accurate definitions and of replicating experiments, and he suggested that scientists should avoid inferring complex mental attributes in animals where such abilities are unproven. In the field of comparative psychology, the rule that complex cognitive processes should not be assumed when simpler processes would suffice became known as 'Morgan's canon'.

Morgan's writings were highly influential in establishing a rigorous science of animal behaviour. His canon later became a cornerstone of both the behaviourist movement and the argument that simple learning processes could underlie seemingly complex behaviour. Ironically, Morgan did not seek to encourage parsimony with his rule, but rather the restriction of behavioural explanations to psychological processes known to occur in that species; in this respect, Morgan's canon 'backfired' (Costall, 1993), as researchers commonly used it to reject the idea that complex mental processes underpinned an observed behaviour. It was Morgan who formulated the definition of instinctive behaviour that is widely held today: an instinct is a behaviour that is entirely determined by the congenital organization of the nervous system. In today's terms, Morgan argued that the only aspects of behaviour that an individual animal inherits are instincts, reflexes, and the capacity to learn.

The study of animal learning was starting to become a respectable subject of scientific study within both psychology and zoology, with emphasis being placed on detailed observation and experimentation. Yet psychology was still immersed in the idea that 'instincts' reigned, and distorted views of evolution continued to influence scientific thinking within this field into the early twentieth century. For example, the American psychologist Granville Stanley Hall, who was one of the founders of psychology as an academic subject, stated that children and adolescents progress through a linear series of 'primitive' mental states before finally reaching a civilized state in adulthood (Hall, 1904). This view of child development, which relied heavily on the ideas of Lamarck and Haeckel, entered popular culture and partly explained the rise of summer camps in North America. These camps were thought to provide boys with the opportunity to express their savage tendencies through appropriate outdoor activities, such as building fires and making tents, while girls were instead expected to be preparing for their roles as wives and mothers.

Sigmund Freud's theories of psychopathology were also greatly influenced by Darwin and Haeckel (Richards, 1987). Freud took Darwin's ideas of sexual selection and the 'instinct' to mate and used them to develop his concept of the libido, the package of largely sexual urges that were the unstated driving force behind human behaviour. Freud's views that one could gauge the inner workings of the human mind indirectly through what was happening on the surface, and that illnesses might be ascribed to forgotten experiences, were influenced by Darwin's work on the expression of the emotions. In addition, Freud's psychosexual theory drew directly from Haeckel's, now discredited, biogenetic law (Sulloway, 1979). If animals at the equivalent developmental stage are sexual creatures, Freud reasoned that infant humans must be too, going through an oral stage when they gain sexual pleasure from the mouth, later to be followed by anal and phallic stages.

One influential psychologist who challenged the Spencerian view of psychology was the American William James, older brother of the author Henry James. William James had become familiar with the advancing field of physiological psychology through his frequent trips to Europe. Initially an admirer of Spencer, James became dissatisfied with the deterministic view of human behaviour that dominated psychology (Plotkin, 2004). Human beings were being caricatured as passive organisms, responding slavishly to changes in the environment. Instead, James reverted to a more Darwinian perspective and proposed that the mind generated ideas (variation) rather than being shaped by the external world, and that the ideas that provided the best way of dealing with the world will be retained (selected). He believed in the importance of adaptation in explaining key features of the mind such as consciousness. James' textbook *Principles of Psychology*, first published in 1890, ran to several editions. The chapter headings included perception, sensation, reasoning, emotion, and memory, and the content was highly materialistic; for example, habits were described as resulting from rearrangements of pathways through the nerve centres. A Darwinian perspective pervaded all of James' work although without actually leading to any specific hypotheses or empirical predictions being made.

Another psychologist whose views on evolution failed to have the impact that they probably deserved was James Mark Baldwin (Plotkin, 2004). Architect of the 'Baldwin effect' and founder of the premier psychological journal, the *Psychological Review*, Baldwin adopted an evolutionary approach to psychology but rejected simple hereditarian views of human behaviour (Boakes, 1984). Baldwin endeavoured to develop psychological principles consistent with evolutionary theory that nonetheless accounted for the influence of cultural inheritance. His major interest was child development, a topic that Baldwin believed had been distorted by the Lamarckian approach of Hall. Through careful observation, he charted the gradual appearance of mental powers in human infants, determined the sequence in which they emerged, and emphasized the possible importance of the social transmission of knowledge in mental development. In many ways, Baldwin was ahead of his time in distinguishing the relatively fixed stages of cognitive development from the role of the individual's developmental experiences and also the interactions between the two. Unfortunately, he was forced to resign from Johns Hopkins University in 1909 after being arrested in a brothel. While Baldwin proclaimed his innocence, his abrasive style had won him few friends in academic circles, and his contributions to psychology were largely written out of the history books. Nonetheless, after moving to Paris, Baldwin continued to impact psychology, particularly through his influence on the Swiss child psychologist Jean Piaget.

William McDougall, an eminent Harvard professor who published *An Introduction to Social Psychology* in 1908, was another strong advocate of human instincts and an avid eugenicist (Sussman, 2014). McDougall argued that human beings possess just seven instincts, including fear, disgust, anger, elation, and parental emotions, with no specific evidence to support this position. However, reaction to any instinct-based theories of human behaviour and eugenics gathered momentum in the early twentieth century. Part of this dissatisfaction was that the concept of instinct was increasingly criticized as being vague and unscientific. More worryingly, one review reported that in the previous twenty years nearly 6,000 types of instinct had been proposed, including the instinct of girls to pat and arrange their hair, and the desire of men to liberate the Christian subjects of the Sultan (Boakes, 1984). Despite growing concerns about the concept of instincts, the prevalence of strongly nativistic explanations for human behaviour was having important and disturbing social ramifications.

At the beginning of the First World War, the US army had allowed the psychologist Robert Yerkes to carry out intelligence testing on the armed forces with a view to improving the intake and efficiency of recruits, and nearly two million men were tested. After the war, when the test results were analysed, the assumptions made by Yerkes and colleagues were strongly hereditarian. The findings, published in *A Study of American Intelligence* by Princeton psychologist Carl Brigham (1923), suggested that intelligence varied with race and, among immigrants, those that had most recently moved to the United States performed worse than those of families with longer residence in the country. These data were taken as proof that, as widely feared, the mental calibre of immigrants was steadily declining. A more likely explanation is that, as immigrants would become increasingly familiar with American culture over time, those immigrants with longest residency would score better on these biased, knowledge-based tests, but this explanation was ignored. Although in the 1920s and 1930s the use of intelligence testing was increasingly criticized, President Coolidge was among those who accepted Yerkes's conclusions, and as a result he imposed the 1924 Immigration Act that restricted immigration to favoured races and nationalities. When signing the Act, Coolidge stated 'America must be kept American'. Fifty years later, the restrictive and racist immigration laws were to be cited by the critics of human sociobiology as a prime example of the dangers of evolutionary methods being applied to human behaviour.

In the first few decades of the twentieth century, the field of psychology experienced a dramatic shift in emphasis towards studying only those behaviour patterns that could be observed and measured. The predictability and control of behaviour patterns, such as reflex actions and stimulus–response learning, became the focus of scientific attention, with the study of learning being the central theme. This school of thought, known as *behaviourism*, began with the publication of work by John Watson (1913). Watson rejected the notion that inheritance played any meaningful part in explaining human behaviour. He stated that we need consider only what is learned to understand human behaviour and that learning is the proper focus for psychological research. In a well-known quotation, Watson boldly claimed:

Give me a dozen healthy infants, well-formed, and my own specified world to bring them up in and I'll guarantee to take anyone at random and train him to become any type of specialist I might select—doctor, lawyer, artist, merchant-chief and, yes, even beggar-man and thief, regardless of his talents, penchants, tendencies, abilities, vocations, and race of his ancestors. (1924, p. 82)

Behaviourist psychology in the United States conformed better with the emerging political ideology that stressed equality of opportunity. A parallel movement had developed in Russia based on the research of the physiologist Ivan Petrovich Pavlov. Lenin is said to have paid a secret visit to Pavlov's laboratory in 1919 to find out whether Pavlov could help the Bolsheviks control human behaviour (Bateson and Martin, 1999). Pavlov told him that 'natural instincts' could be abolished by a form of learning now known as 'Pavlovian conditioning', a view so congenial to Lenin that it became the party line, and Pavlov's research was widely promoted. By the 1930s, the idea of 'instincts' had largely disappeared from experimental psychology.

Shortly after the rise of behaviourism in psychology, a similar reaction against instinct and hereditarian views occurred within anthropology (Plotkin, 2004). The leader of the new movement was Franz Boas. In 1883, as a 25-year-old student from Germany, Boas went to live among the people of Baffin Island in northern Canada and became aware of a relative arbitrariness to human customs (King, 2019). This impression was strengthened by expeditions to study the indigenous populations of British Columbia, where he observed extensive variation in both cultural practices and artefacts over relatively short geographical distances. Boas dismissed the idea that all cultures progress through similar stages and emphasized instead the importance of local history and chance occurrences. He urged anthropologists to engage in the careful study of individual communities, as exemplified by the detailed documentation of Native American languages and dialects that Boas carried out in collaboration with fellow anthropologist Ella Cara Deloria. Boas advocated against using overarching generalizations and universal rules when describing human cultures.

Boas and his students founded the new field of *cultural anthropology*, which emphasized the pre-eminent role of context and social history. Rather than studying museum collections of artefacts, this new wave of anthropologists attempted to gain first-hand knowledge of specific communities (King, 2019). For instance, Margaret Mead, who was to become one of the most famous anthropologists of the twentieth century, travelled to Polynesia and collected personal stories from adolescent girls, which were summarized in her best-selling book, *Coming of Age in Samoa* (1928). Mead concluded that the social turmoil experienced by many American adolescents was not a universal or inevitable stage of life, but instead resulted from particular social constraints and strictures. Mead refused to accept the superiority of Western culture and described the wide variety of social rules, for instance regarding sex and relationship, that she observed in different communities as 'experiments in what could be done with human nature' (Mead, 1930, p. 2). Ruth Benedict, another of Boas' students who would eventually become President of the American Anthropological Association, studied the folktales of American Southwestern indigenous populations, where women held property rights and cross-dressing was accepted. In *Patterns of Culture*, Benedict (1934) argued that different cultures represent equally valid ways of life and that everyone should practise tolerance and reflect on how their personal viewpoints are patterned by their own cultural upbringings.

The fields of psychology and anthropology were each grappling with the question of 'instincts' in their own way, and, in both cases, evolution was the baby that went out with the bath water. Ironically, psychology, anthropology, and the other human sciences rejected evolution at precisely the time that evolutionary theory was really coming together. The modern synthetic theory of evolution was forged in the 1930s, with the integration of Mendel's genetics and Darwinian thought, the rejection of Lamarckian inheritance, and with natural selection re-established as the major evolutionary process. The classic works of Theodore Dobzhansky (*Genetics and the Origin of Species*, 1937), Ernst Mayr (*Systematics and the Origin of Species*, 1942), Julian Huxley (*Evolution: The Modern Synthesis*, 1942), and George Simpson (*Tempo and Mode in Evolution*, 1944) showed how this new *Modern Synthesis* could be employed to make sense of evolutionary lineages and of the characters of contemporary populations of organisms. Evolutionary theory gained a solid theoretical foundation through the works of Ronald Fisher, John Haldane, and Sewell Wright in the 1920s to 1950s, in which the methods of population genetics and the mathematical theory of evolution were worked out, and key concepts such as fitness were defined. Evolutionary biology could now be regarded as a mature science, although it continued to be associated with eugenicist views. For example, Fisher's influential *Genetical Theory of Natural Selection* (1930), contains five chapters on eugenics, including sections on 'The decay of the ruling classes', 'Contrast with barbarian societies', and 'Social promotion of fertility'.

One scientific development that resulted from the emergence of the modern synthetic theory of evolution and the advances in genetics was the cataloguing of genetic variation in human populations. This research soon revealed that the amount of genetic variation between human populations is extremely small compared with the amount of genetic variation within populations (Lewontin, 1972). This finding, which has been replicated using more recent data (Rosenberg et al., 2002), supported the vigorous arguments that many evolutionary biologists were making against racism. By the end of the twentieth century, new methods in genetics, including the sequencing of modern and ancient human genomes, allowed researchers to gain a much clearer understanding of the history of our species. Such data have revealed that all human beings are extremely closely related to each other and share an African origin. To put this statement into perspective, the amount of genetic variation between humans across the globe is far smaller than the genetic variation found between groups of chimpanzees in central Africa (Bowden et al., 2012). Additionally, those human traits that are known to be influenced by genes, such as skin colour and eye shape, vary on a continuous basis within and across regions, which contradicts that idea that people can be divided into discrete 'races' (Graves and Goodman, 2021). The accumulated empirical evidence means that 'scientific racism' does not deserve a place in any modern academic discipline.

Ethology and the resurrection of instinct

As the majority of psychologists and anthropologists disregarded evolutionary arguments, an increasingly persuasive body of knowledge and valuable set of methodologies for the study of behaviour were being developed. This science became known as *ethology*, from the Greek 'ethos' meaning character. The ethologists made a significant and lasting impression on the field of zoology by drawing attention away from preserved museum specimens and mounted skeletons to the study of animals behaving in their natural settings. Using a knowledge of natural history, the ethologists set out to examine the robust behaviour patterns that are seen within one species and not another. The idea of 'instinctive' behaviour was once again re-emerging. Here, we spend some time describing ethology, as this work provides much of the background for the fields that we introduce in the rest of the book. In the subsequent section, we describe some of the work on human behaviour that was carried out under the name of ethology, some of which was damaging to its scientific reputation.

In the late nineteenth and early twentieth century, two scientists, Oskar Heinroth in Germany and Charles Otis Whitman in the United States, were independently documenting patterns of movements, such as courtship behaviour in birds (Burkhardt, 2005). Heinroth observed that the precise movements of ducks engaged in courtship was highly characteristic of a species, and that the similarities and differences between species could be used in exactly the same way as physical characteristics to trace common ancestry and reconstruct the evolutionary past. Whitman's painstaking studies of the reproductive behaviour of pigeons led to similar conclusions, while African American entomologist Charles Henry Turner conducted relevant pioneering studies of insect behaviour. A few decades later, a young Austrian anatomy student called Konrad Lorenz, heavily influenced by Heinroth's work, came to the same conclusion that the methods employed in comparative morphology could be applied to the behaviour of animals. Lorenz was determined that 'the phylogenetic view' (his term for an evolutionary perspective) should triumph in the study of animal behaviour. From early childhood, Lorenz had an inordinate love of animals and, knowing nothing of Spalding's work, had independently discovered imprinting through his experiences in hand-raising flocks of geese in his home village of Altenberg. The picture of Lorenz being followed around the Austrian countryside by a line of young goslings has become one of the most enduring images of ethology.

In 1936, Lorenz met Nikolaas Tinbergen, a zoologist at the University of Leiden in the Netherlands, who had developed a research programme characterized by the observational and experimental study of animals in their natural habitats (Kruuk, 2004). They were amazed at the similarities of their views and struck up an immediate friendship. Lorenz and Tinbergen were the true founders of ethology and pioneered a novel approach to the study of behaviour. By the early 1950s, ethology had emerged as a new discipline, with Lorenz as its father figure and *The Study of Instinct* (1951) by Tinbergen its classic text (Thorpe, 1979). The elegant studies of another Austrian ethologist, Karl von Frisch, on the 'waggle-dance' communication in the honeybee are arguably one of ethology's most famous insights. Ethology also flourished in Britain, with William Thorpe and Robert Hinde at the University of Cambridge pioneering the study of birdsong and behavioural development in birds and primates, and at the University of Oxford, where Tinbergen moved in 1950. Ethology also made an impact in the United States by the middle of the century (Dewsbury, 1995), particularly through the work of William Morton Wheeler and Karl Spencer Lashley, with international travel and conferences providing new opportunities for the exchange of ideas.

The ethological method typically began with an extensive period of observation of the animal in its native environment, followed by a careful description of the relevant behaviour patterns, known as an 'ethogram'. A variety of stereotypical behaviour patterns or *fixed motor patterns* for that species were identified, such as the courtship displays of ducks or web-making behaviour of spiders. The ethologists also described behaviour patterns that arise during development prior to having an apparent function; for example, a newly born duckling will dab its beak onto the oil gland at the base of the tail during grooming before the gland is functioning to produce oil (Lorenz, 1965). Such behaviour patterns are ostensibly inconsistent with an explanation based on reinforcement learning, as are behaviour patterns that require little practice. For instance, many young birds can fly at their first attempt. To the ethologists, 'instinct' was an inherited and adapted system of coordination within the nervous system. Tinbergen, in particular, focused on the survival value of specific behaviour patterns and was exemplary at designing simple experiments that would test causative and functional hypotheses in natural conditions. For example, Tinbergen examined how female digger wasps (*Philanthus triangulum*) locate the entrance to their nests. These wasps dig small holes in the soil in areas of open ground, and Tinbergen and his students would place a ring of pinecones around a nest entrance. The wasp would leave the nest to forage, first spending a few seconds circling around the site, and the researchers would then move the pinecones to a location nearby and observe the wasp initially searching for the nest entrance in the incorrect location. They concluded that the wasps used visual cues in memory formation (Tinbergen and Kruyt, 1932).

Some ethologists tried to explain instincts in physiological terms, in a manner that was subject to experimental investigation, with particular focus on the role of hormones in eliciting behavioural responses (Hinde, 1966). Ethologists also attempted to produce models to explain how signals in the environment, including those produced via social interactions, result in appropriate behavioural responses. For instance, Lorenz suggested that the tendency to produce instinctive behaviour builds up over time and, when activated by the appropriate stimulus, finds expression in a fixed motor pattern. In his *hydraulic model* of behavioural drives, which drew heavily on the earlier work of experimental psychologist Wallace Craig, Lorenz (1950) suggested that the drive to feed, fight, or mate gradually built up over time through the accumulation of 'action-specific energy', in a manner analogous to the build-up of liquid dripping into a tank. In the presence of a signal, a releasing mechanism would act like a valve at the bottom of the tank, and the escaping liquid would represent the expression of the behaviour pattern. Lorenz also argued that, when the build-up of energy became so high, a motor response would be produced even in the absence of the appropriate stimulus. The ethologists attempted to arrange the releasing mechanisms into hierarchical systems, to compare the strengths of different drives and to measure the thresholds at which behavioural responses were elicited. However, hydraulic models received criticism from other ethologists for failing to fit with all of the observed data (Hinde, 1956). For instance, in some cases, the presence of a rewarding stimulus could elicit a specific behaviour pattern and then increase, rather than decrease, the likelihood that the same behaviour pattern would be exhibited again.

More broadly, ethology was constantly engaged in a running battle with the North American school of *comparative psychology* that arose from behaviourism. The two groups of researchers shared an interest in animal and human behaviour, but they approached the topic from very different viewpoints. The ethologists, who worked largely in Europe, were mostly biologists and naturalists, and they largely studied animals in their natural habitats. In contrast, the comparative psychologists studied animal behaviour in laboratory settings and, despite their name, were not concerned with comparisons between species. They tended to focus on just one or two species, such as rats or pigeons, because they believed that the general rules of behaviour would hold across species, regardless of the experimental context. For instance, the comparative psychologists studied how exposure to certain stimuli produced specific responses that could become conditioned 'habits'. They rejected the idea that behaviour could instead be explained in terms of motivations or 'drives', suggesting that these terms were vague and relied on inferring mental states that could not be measured. In response, the ethologists maintained that the so-called general rules emphasized by comparative psychologists were artefacts of the impoverished experimental conditions. They were also concerned that comparative psychologists failed to take into account how these rules were employed by animals in real-life settings, which often involve interacting with other individuals.

An important critique of ethology, and specifically the concept of 'innate' behaviour, was written by the American psychologist Daniel Lehrman in 1953. Lehrman dismissed the ethologists' accounts of 'innate' behaviour, first, because organisms never develop in complete isolation from their environment and therefore one could never know that a behaviour pattern was uninfluenced by external events, and second, because 'innate' was defined in terms of excluding what is learned, it would never be a usable concept, given that learning can rarely be discounted. Lehrman also pointed out that behaviour patterns that are universal to a particular species are not necessarily 'innate', as all individuals could instead be exposed to the same relevant stimuli or conditions during early development and thereby exhibit the same trait in later life as a result of those early experiences. Another influential comparative psychologist, Theodore Schneirla (1966), also convincingly argued that the relative importance of 'innate' and 'acquired' effects on behaviour patterns could not be completely separated and that an individual's development is a complex interaction of genetic information, the developing organism, and its environment. However, the false distinction between 'innate' and learned behaviour continues to be found to this day.

Perhaps because the ethologists were so preoccupied with their battle with the comparative psychologists, they constantly stressed the characteristic fixed behaviour patterns of a species and neglected how individuals vary within a species. That variation was central to Darwin's perspective on evolution and to the process of natural selection. Lorenz's early training in comparative anatomy may have accounted for his tendency to focus on typical specimens or 'types', and his influence may help to explain why many ethologists repeatedly made the mistake of thinking that natural selection was a process that operated 'for the good of the species'. If all individuals are thought to behave in the same manner, then it is easy to envisage that their interests are aligned. Eventually, the ethologists conceded that 'instinct' was not an adequate explanation for animal behaviour, not least because it discouraged consideration of how behaviour develops (Hinde, 1982). Their own experiments eventually led ethologists to the realization that concepts such as 'innate' and 'instinct' were slippery and vague. As later argued by Patrick Bateson (1996), these terms possess several meanings, including present at hatching or birth, unchanging throughout development, shared by all members of a species, adapted over the course of evolution, and a behavioural difference caused by a genetic difference and not learned.

This plurality of definition might have proven workable had the different qualities attributed to instinct always co-occurred, as Lorenz originally envisaged. However, the ethologists' studies revealed this assumption was incorrect. For instance, Gilbert Gottlieb (1971) showed that the preference of a newly hatched duckling for the maternal calls of its own species is affected by hearing its own vocalizations in the egg; this preference is a characteristic that is present at hatching but not unlearned. Tinbergen (1953) showed that a newly hatched gull chick will immediately peck at its parent's bill to initiate feeding but established that the accuracy of pecking improves after hatching; this bill-pecking behaviour is therefore species-typical and present at hatching, but nonetheless modified during early life. Experimental studies were also revealing the constant interplay between internal and external factors during development. For instance, Lehrman (1965) discovered that the male dove's courtship dance triggers the internal change in the female dove's hormones that precede mating and reproduction. This realization led Tinbergen (1963) to add '*How does a behaviour develop?*' to the three questions of biology that had previously been outlined by Julian Huxley; these questions were what are (1) proximate physiological causes, (2) the function (or survival value), and (3) the phylogenetic (or evolutionary) history of a behaviour pattern? Ethology had identified four important classes of question that can be asked about behaviour, and these questions still form the core of modern animal behaviour research.

One of Lorenz's major contributions to the understanding of animal behaviour was his view that learning itself is an evolved ability, and that both instinct and learning are of importance and not mutually exclusive. In *Evolution and Modification of Behavior*, Lorenz (1965) provided a partial solution to the nature–nurture debate that has generally been overlooked; he introduced the concept of the 'innate school-marm', that is, an evolved internal mechanism that instructs learning. One of the most important contributions of ethology to the social sciences is the idea that the development of an individual is channelled but not predetermined, with evolved predispositions influencing when, what, and how an animal learns. The study of non-human animals has provided important insights into the mechanism, development, and evolution of behaviour, and all of the contemporary evolutionary approaches to human behaviour that we review in this book build on the foundational work of the ethologists and comparative psychologists to some extent. In Chapter 8, we briefly examine how animal behaviour research continues to play a role in expanding our understanding of behaviour and cognition.

Human ethology

In 1973, Lorenz, Tinbergen, and von Frisch were awarded the Nobel Prize for Physiology or Medicine 'for their discoveries concerning the organization and elicitation of individual and social behaviour patterns'. That the first Nobel Prize to be awarded for the study of behaviour and the causes of behaviour went to ethologists caused great discussion and dispute among the comparative psychologists. However, the award reflected the optimism current at the time that work in ethology would generate new understanding in related fields, such as medicine and psychiatry, and shed light on human behaviour.

From the outset, Lorenz believed that ethology would provide important insights into human behaviour. In virtually all of his popular books, the final chapter revealed what the preceding pages had to say about humans. However, his views on human behaviour were tarnished by his political entanglements (Kalikow, 2020). In the early 1940s, Lorenz wrote thinly veiled scientific papers that are commonly interpreted as supporting the Nazis, their ideal of racial purity and the selecting out of so-called degenerate elements in society. Many years later, Lorenz confessed that he had found some of the

Nazi theories attractive but had been politically naïve and had no conception that they would result in genocide (Evans, 1975). It was only late into the war that he realized the evil of Nazism. Nonetheless, the reader of these articles will not find it difficult to understand why Lorenz's critics would charge him with abusing biological arguments to justify racism. In contrast, Tinbergen's experiences in occupied Holland and in a hostage camp, left him unable even to bear the sound of spoken German (Robert Hinde, personal communication). The Second World War delayed the development of ethology, cutting off relationships between colleagues and friends, although Lorenz and Tinbergen were to renew their friendship a few years later.

Lorenz's 1966 book *On Aggression* caused a major furore, and greatly upset many intellectuals and social scientists (Montagu, 1968). Lorenz argued that fighting and war are the natural expression of 'instinctive' human aggression, which, according to his theory, inevitably welled up in human beings, unless otherwise expressed, and would be discharged spontaneously and without reason. Despite a final avowal of optimism, Lorenz painted a bleak picture:

> An unprejudiced observer from another planet, looking on man as he is today, in his hand the atom bomb, the product of his intelligence, in his heart the aggressive drive inherited from his anthropoid ancestors, which this same intelligence cannot control, would not prophesy long life for the species. (1966, p. 40)

In typically blunt terms, Lorenz suggested that any attempts to eliminate aggression by appropriate training or by shielding human beings from all circumstances that might elicit it 'have no hope of success whatever' (1966, p. 239). Lorenz argued that the only chance for humanity was to face up to the grim reality, charging us 'know thyself', and he suggested one or two rather uncompelling solutions, such as to engage in more sport to release aggressive urges.

Lorenz's book provoked considerable hostility and was disowned by many English-speaking ethologists (Salzen, 1996). Critics objected to his extrapolation from animals to humans, many argued that aggressive behaviour was learned, and others drew the disturbing conclusion that if aggression was the expression of an inescapable urge, then war is unavoidable. The opposition and debate continued for more than twenty years. In 1983, a group of expert scientists conferred at a meeting on aggression in Spain and drew up what has become known as 'The Seville Statement on Violence' (Table 2.3). Endorsed by major professional bodies and published in prestigious journals, the statement was eventually adopted and disseminated by UNESCO, with the express purpose 'to dispel the widespread belief that human beings are inevitably

Table 2.3 The 1986 Seville Statement on Violence

1.	It is scientifically incorrect to say that we have inherited a tendency to make war from our animal ancestors.
2.	It is scientifically incorrect to say that war or any other violent behaviour is genetically programmed into our human nature.
3.	It is scientifically incorrect to say that in the course of evolution there has been a selection for aggressive behaviour more than for other kinds of behaviour.
4.	It is scientifically incorrect to say that humans have a 'violent brain'.
5.	It is scientifically incorrect to say that war is caused by 'instinct' or any single motivation.

disposed to war'. Ironically, Lorenz makes none of the 'scientifically incorrect' statements of the Seville tract, although he comes close to it.

Lorenz was far from the only ethologist to address human behaviour. An entire sub-discipline of 'human ethology' emerged in due course. For example, in his retirement, Tinbergen controversially applied ethological methods to the study of childhood autism, in collaboration with his wife Elisabeth Tinbergen. He also considered the possibility that the evolved traits of humans left them poorly suited to rapidly changing, modern environments (Vicedo, 2018). The British psychiatrist, John Bowlby, developed this idea further. Greatly influenced by Robert Hinde, Bowlby adopted an ethological perspective to help explain why young children become attached to their mothers and why they experience great anxiety when deprived of this contact. Lorenz's student, Irenäus Eibl-Eibesfeldt, extended Darwin's study of emotion by travelling round the world photographing the facial expressions of people from different communities expressing particular emotions, including indigenous populations with little previous contact with Westerners. Paul Ekman carried out similar studies, each concluding that the same facial expression represents the same feeling all round the world, although subsequent research has shown that how emotions are felt and expressed is also influenced by cultural norms (Corado et al., 2018). However, human ethology did not live up to its early promise, perhaps because many ethologists themselves recognized the need to take account of the peculiar complexities of human beings (Hinde, 1982, 1987).

The scientific credibility of applying ethological methods to studying human behaviour was particularly damaged by the popularized version put forward by Desmond Morris, a British zoologist and curator of mammals at London Zoo. In 1967, Morris created an even bigger controversy with the publication of *The Naked Ape* than Lorenz had with *On Aggression*. It was an extraordinarily popular book that was to sell over ten million copies and be translated into every major language. The basic premise was that humans can best be understood as typical primates that had turned to hunting. 'His whole body, his way of life, was geared to a forest existence, and then suddenly... he was jettisoned into a world where he could survive only if he began to live like a brainy, weapon-toting wolf' (1967, p. 16). Morris argued that 'the fundamental patterns of behaviour laid down in our early days as hunting apes still shine through all our affairs' (1967, p. 26), and he went on to provide unsupported evolutionary explanations for our parental behaviour, sexual behaviour, aggression, and virtually every other aspect of our daily lives.

Morris depicted himself as a simple ethologist describing the human animal in honest zoological terms, giving readers straight biological truths about the animal selves they had been loath to contemplate. However, the flowing prose was rife with sex and sensationalism, and he frequently touched on sensitive topics and drew wildly speculative conclusions. For instance, Morris (1967) stated that pornography and prostitution are comparatively harmless and may actually 'help', that women are wrong to stop their husbands 'going out with the boys', and warned that, if women take on masculine traits, they risk making their sons homosexual. Many fellow ethologists understandably disapproved of Morris' writings; for example, Lorenz complained that Morris treated humans as if culture was a biologically irrelevant phenomenon (Evans, 1975). In the 1960s and early 1970s, there was a proliferation of popular books that set out to explain aspects of current human behaviour as reflections of our evolutionary past, such as Robert Ardrey's (1966) *The Territorial Imperative*, Lionel Tiger's (1969) *Men in Groups*, and Tiger and Robin Fox's (1971) *The Imperial Animal*. Commonly, these books excited controversy as the descriptions of purported 'innate' behavioural tendencies were seen as justifications for existing social inequalities (Segerstråle, 2000).

A history of sense and nonsense

With hindsight, we can now see that books like *The Naked Ape* are representative of a long line of texts, dating back to those of Darwin's contemporaries, that use evolutionary arguments to tell the reader what is 'right', 'natural', or 'inevitable'. From the beginning, self-appointed evolutionary evangelists have been serving up biological 'home truths', while others, such as Thomas Huxley, have objected to the more excessive claims and suggested that prejudice and ulterior motives lie behind their conjectures. Little wonder, then, that many people are wary of evolutionary arguments.

We can also see that, historically, certain ideas have tended to go together: a Lamarckian view of evolution, with species arranged on a ladder and a linear, progressive concept of change, perhaps inevitably engenders prejudice as some evolved forms must be regarded as more advanced, or 'higher', than others. Many of the inequitable views on human races indirectly resulted from this Lamarckian viewpoint. In contrast, the Darwinian conception of evolution stresses within-species variation and rejects the typological thinking that is inherent in racism. In addition to the role of natural selection, modern Darwinism places considerable emphasis on chance events such as mutation and genetic drift, and natural selection does not necessarily result in the progression of populations towards an end goal or 'higher' state. In fact, the misrepresentation of evolution as progressive was so apparent to Darwin that, in his notebooks, he reminded himself to 'never say higher or lower' (Gruber, 1974), sound advice that unfortunately even he did not always follow. Evolutionary biologists now recognize that it is impossible to define any non-arbitrary criteria by which progress in evolution can be measured. As no variant can be regarded as more advanced than others, Darwinian evolution is inconsistent with social Darwinism and with attempts to use scientific arguments to promote racism. It is largely by distorting Darwinian thinking that evolution has been used to justify prejudice and inequality. Most of the negative features sometimes unfairly attributed to evolution, including racism, sexism, genetic determinism, and social Darwinism, do not come from Darwin but from others who distorted his theory.

Another characteristic of Darwin's work from which we can learn was his care and diligence in accumulating as much evidence as possible on the subject of his investigations. He finally published *The Origin of Species* twenty years after the idea of natural selection had first sparked his imagination, and a further decade passed before Darwin said anything substantive about human evolution. Darwin's books and scientific papers are overflowing with evidence and illustrative examples, which are painstakingly weighed up in support of his hypotheses and counter to alternative explanations. This careful approach may be contrasted with other works that we have mentioned in this chapter, which make bold statements based on little or no supportive evidence. By the end of this book, we will see that the most compelling evolutionary explanations of human behaviour are those backed up by rigorous accumulation of data, ideally from a large number of sources. Darwin was also highly aware of how society would respond to his work and its possible implications, and he took care to build a water-tight case before making his views known to the world.

Although Darwin was not always right in every respect, his idea of evolution by natural selection has withstood the test of time. Over the years, evolutionary reasoning has made invaluable contributions to understanding of topics such as the relationship between learned and inherited traits, the causes of individual differences, and the development of behaviour. It has also led to the rejection of both genetic determinism and the *tabula rasa* view that human behaviour is infinitely malleable. The investigations of the biological basis of imprinting and birdsong by ethologists have been at the forefront of research into learning and have led to a new comprehension of how behavioural aspects of development can be linked to an understanding of brain mechanisms. Such research strongly implies that there is no meaningful nature–nurture dichotomy. Genes take their cues from the environment, and learning relies on gene expression. In addition, evolutionary research has contributed to the debate against racism by showing that genetic variation within human populations swamps differences between populations. Yet, for many, such achievements are overshadowed by the negative uses of evolutionary reasoning.

Recent times have furnished fresh evolutionary insights and new methods that, if used correctly, promise to lend a new impetus to the quest to understand human behaviour and society. These ideas will be introduced over the next few chapters. We begin with the sociobiological revolution, where we see that the controversies that surrounded human ethology were nothing compared to the fracas over human sociobiology.

3
Human sociobiology

When Konrad Lorenz, Nikolaas Tinbergen, and Karl von Frisch collected the Nobel Prize for their contributions to the study of animal behaviour in 1973, the field of ethology was already starting to be overshadowed by the rise of a new perspective within evolutionary biology. The new approach, known as *sociobiology*, built on the work of the ethologists but laid more emphasis on the functional significance of behaviour (questioning why animals have been selected to behave in particular ways) at the expense of causal processes (e.g. investigating what stimuli elicit specific behaviour patterns). Sociobiology brought with it a range of novel methods and insights that initiated a radical overhaul of evolutionary thinking in the context of animal behaviour. While in Britain and the rest of Europe the transition from ethology to sociobiology may have been more gradual, the new field took off suddenly in North America, perhaps because ethology was less prominent there. By the spring of 1976, entire courses were being offered on sociobiology at major universities in the United States, and, by the end of the decade, several new scientific journals concerned with sociobiological issues had been created. All of a sudden, eager researchers had a fresh methodology, a new set of questions, and the spring of optimism in their step. Imagine the excitement. Puzzles that had taxed the minds of great thinkers, including Charles Darwin, just seemed to be coming into focus under the powerful resolution of sociobiology's tools. Why then, when the behaviour of ants, gulls, and monkeys seemed to fall suddenly into place, should these new methods not be applied to our own species?

The pioneers of this new way of thinking were George C. Williams, William Hamilton, Robert Trivers, and John Maynard Smith. However, two books brought the attention of the general public to the ideas behind sociobiology. In 1975, Harvard professor Edward O. Wilson published *Sociobiology: The New Synthesis*, which made an immediate impression and resulted in a storm of controversy soon after publication. Wilson's important contribution was to create and name the field of 'sociobiology', by showing its scattered practitioners that it existed, and to demonstrate its feasibility and importance (Segerstråle, 2000). A year later, Oxford zoologist Richard Dawkins brought out *The Selfish Gene*, arguably the most popular scientific book of the twentieth century. These books were a celebration of the 'gene's-eye view', the notion that, if we wish to understand what characters ought to evolve, it is a convenient and useful heuristic to look at the problem from the perspective of the gene and ask which traits would be most likely to increase its frequency in the next generation. Both books successfully captured the potential and excitement provided by the novel ideas and methods that collectively had reinvigorated evolutionary biology, and it is impossible to overstate their impact. Biologists all round the world started rewriting their lecture courses around these two monographs, and laypeople were able to comprehend complex ideas being discussed in evolutionary biology.

Although Dawkins argued that culture took humans into a new realm, Wilson, a scientist renowned for the courage of his convictions, was certainly not shy of applying sociobiological methods to humans. In the final chapter of his book, Wilson turned his thoughts to human behaviour, offering bold and speculative evolutionary hypotheses for controversial topics such as gender roles, aggression, and religion. He stated quite openly that one of the goals of sociobiology was to 'reformulate the social sciences in a way that draws these subjects into the Modern [evolutionary] Synthesis' (1975a, p. 4). Wilson's book was to catalyse the appearance of a stream of works using and extending the theme of human sociobiology. For other researchers, emboldened by a revolutionary zeal, human behaviour had the appearance of rich, easy pickings. The result was a land rush of biologists into the territory of the human sciences, where they received an extremely hostile reception. The unprecedented tumult over sociobiology was to prove the biggest scientific controversy of the decade.

There was, of course, far more to the development of sociobiology than the matters that have concerned its bearing on humanity. The application of sociobiological methods to the study of animal behaviour proved relatively uncontentious, and this field has since morphed into *behavioural ecology*. Likewise, the use of comparative studies to draw inferences about animal behaviour and cognition has become a rigorous and respected research field. Nonetheless, in this chapter, we will take an anthropocentric look at sociobiology by reviewing its principal ideas and methods, providing examples of the application of these methods to our own species, and then discussing the numerous criticisms that were presented against this research programme.

Key concepts

Wilson described sociobiology as 'the systematic study of the biological basis of all social behaviour' (1975a, p. 4), but this all-encompassing statement captures little more than the breadth of his vision. Wilson synthesized a new discipline by drawing together experimental and theoretical studies of animal demography, population biology, communication, grouping behaviour, parenting, and aggression in species ranging from micro-organisms through invertebrates to birds, mammals, and finally human beings. By 1975, groundbreaking developments in evolutionary theory and ecology had led to a more rigorous theoretical framework for the study of animal behaviour. What set sociobiology apart from ethology was the use of a set of key conceptual tools that are set out below, including the *gene's-eye view*, *kin selection*, and *reciprocal altruism*. *Optimality models* were also particularly central to Wilson's synthesis (these will be described in more detail in Chapter 4); furthermore, *game theory* and *evolutionarily stable strategies* received considerable attention and provided new insights into the evolution of social behaviour.

Some of the advances arose in response to the idea of 'group selection', which had become the standard evolutionary explanation for behaviour. Most ethologists had not questioned the concept of group selection, and many presumed that individual organisms were selected to behave for the good of the group or the species. With the advent of sociobiology as a discipline, greater attention was being paid to whether selection was instead acting predominantly at the level of the individual organism. The innovative arguments set out against group selectionist thinking were to lead to important advances in the study of animal behaviour. In this section, we present an introduction to some of the ideas and methods of sociobiology, and we illustrate how each was applied in a striking and controversial way to interpret our own species.

The gene's-eye view

Advocates of group selection had maintained that many aspects of the social behaviour of animals could be explained by the idea that animals made sacrifices for the good of the group. For instance, some ethologists had suggested that animals would forgo mating or even cause their own deaths in an attempt to limit their population size, thereby avoiding overexploitation of their food supplies that might lead to a population crash. This view of evolution was most forcefully advocated by a Scottish ecologist, Vero C. Wynne-Edwards, in his book *Animal Dispersion in Relation to Social Behaviour* (1962). Wynne-Edwards argued that limitation of population growth could be achieved by some individuals altruistically restraining their reproduction and thereby provided an explanation for why subordinate individuals within populations often do not breed. Under such circumstances, groups of individuals or species that limited their reproduction might be more likely to thrive than groups or species that overexploited their habitats. Animals were assumed to use vocalizations, displays, and aggregations to assess population density and use this information to decide whether or not to reproduce. Similarly, in *On Aggression* (1966), Lorenz described highly restrained and ritualized disputes between animals, with such fights taking place according to some equivalent of the Queensberry rules that govern boxing. He argued that these encounters should be seen as competition between individuals to determine who had earned the right to breed and who should withdraw, leading to contracts that would be favourable for the future of the species.

Although these explanations of animal behaviour seemed superficially plausible, the phenomena explained by Wynne-Edwards and others in terms of group selection could instead be explained in terms of individuals attempting to maximize their own reproductive success. In 1964, John Maynard Smith published a short rebuttal of group selection, and, in 1966, David Lack challenged Wynne-Edward's group selectionist interpretations of the empirical evidence, in particular those on bird populations. However, the most powerful platform against group selection was provided by George C. Williams in his classic book *Adaptation and Natural Selection* (1966). Williams was highly dissatisfied with group selection arguments. He pointed out that group selection was unlikely to occur where individuals are able to cheat the system for their own benefit, as such cheaters would out-compete other members of the population and increase in numbers at the expense of others in the group. He also pointed out that the movement of individuals between groups would erode group differences and weaken group selection further. Williams convincingly demonstrated that a simpler and more plausible explanation comes to light if one drops down a further level from the individual and thinks about what characteristics a gene would need to have to increase its representation in the next generation. Williams stated that a 'gene is selected on one basis only, its average effectiveness in producing individuals able to maximize the gene's representation in future generations' (1966, p. 251).

The social behaviour described by Wynne-Edwards in group selection terms, for example, the lack of breeding by individuals in poor condition or low in the social hierarchy, could instead be explained in terms of natural selection acting within groups. For instance, a genetic variant that increased the probability that its carrier would delay breeding if the individual was in poor condition, such that it would only be wasting time and resources, might have a selective advantage over another genetic variant that encouraged such an individual to attempt to breed under all circumstances. Similarly, disputes over territories may be understood as competition for the resources required for breeding, and losers may not be able to breed, or may be better off not attempting to breed, rather than altruistically refraining for the population's sake. Later, this gene-centred perspective was still more powerfully expressed in Dawkins' *The Selfish Gene* (1976) The importance of taking the gene's-eye view will become evident as we discuss the ideas of kin selection, parent–offspring conflict, and reciprocal altruism.

Kin selection

The main difficulty facing evolutionary biologists opposed to group selection was to explain altruism. Why should an individual behave in a way that decreases its own chances of surviving and reproducing, while increasing another individual's reproductive success? How could such apparently self-sacrificial behaviour have evolved? For example, in many colonies of ants, bees, and wasps (the Hymenoptera), the majority of individuals, known as the workers, are not able to reproduce at any point in their lifetimes and instead devote their efforts to raising the offspring of one or more reproductive females, the queens. In *The Origin of Species*, Charles Darwin described the presence of these workers as 'the one special difficulty, which at first appeared to me insuperable, and actually fatal to my whole theory' (1859, p. 257). This conundrum had been puzzling evolutionary biologists for over a century. It was not until 1964 that a British graduate student, William Hamilton, finally devised a satisfactory solution that was consistent with modern genetics: the answer was kinship. As close relatives share copies of many of the same genes, any genes underpinning altruism could potentially increase in frequency in the next generation if individuals carrying those genes helped closely related kin to reproduce.

Hamilton based his work on that of population geneticist Ronald. A. Fisher, who had retired from the Department of Genetics in Cambridge in 1957, around the time that Hamilton started his undergraduate studies at the university (Segerstråle, 2013). Hamilton's lecturers at Cambridge and then London were mainly group selectionists who disregarded Fisher's work, but Hamilton was not dissuaded from pursuing his goal of understanding the evolution of altruism. Another leading British population geneticist, John B. S.

Haldane, had proposed a group selectionist model of altruism, but Hamilton had quickly rejected it. Instead, Hamilton (1964a, 1964b) used mathematical models to show that selection could favour genes that encouraged individuals to behave altruistically towards relatives depending on their degree of relatedness. Hamilton illustrated his results by stating:

> To express the matter more vividly, in the world of our model organisms, whose behaviour is determined strictly by genotype, we expect to find that no one is prepared to sacrifice his life for any single person but that everyone will sacrifice it when he can thereby save more than two brothers, or four half-siblings, or eight first cousins. (1964a, p. 16)

Maynard Smith, who introduced the term 'kin selection' to refer to the evolution of kin-related altruism, caused ructions with Hamilton by suggesting that Haldane had pre-empted his work (Segestrale, 2013). Reputedly, in a pub one evening, Haldane had joked that he would jump into a river to save two brothers or eight cousins, a quip that only makes sense in terms of counting shared genes. Later, Haldane (1955) gave a more precise verbal description of how kinship affects altruism but, crucially, stopped short of developing a mathematical model. Hamilton eventually conceded that Haldane's previous comments on the topic suggested that he had a similar understanding of the importance of degree of relatedness. However, Hamilton's formal theory of inclusive fitness went far beyond Haldane's verbal remarks and undoubtably revolutionized our understanding of animal social behaviour.

The basic idea of *kin selection* is straightforward. Consider the example of an individual that behaves altruistically to a relative at some cost (denoted as c) to its own life prospects, but that the act provides benefits (denoted as b) to a relative in terms of increased chances of survival and reproduction. If the propensity to act altruistically is increased by genes that are also present in that relative, then, although the altruist's chances of passing on those genes directly are decreased, the likelihood that the relative will do so is enhanced. Selection of this behaviour will occur if the fitness cost to the altruist (c) is less than the benefit to the relative (b) multiplied by the probability that the relative possesses the same gene as a result of their relatedness (r). This hypothesis, which can be denoted as $c < br$, became known as Hamilton's rule. Initially, Hamilton struggled to get his theory across, partly due to the reluctance of some academics to countenance the idea of genes influencing human altruism, especially in the decades following the end of the Second World War, but he eventually managed to publish his work. We will see that the importance of weighing up the costs to the donor and benefits to the receiver is central to many of the key concepts in sociobiology.

Kin selection is of particular relevance to the study of social Hymenoptera (such as bees, wasps, and ants) because of their unusual form of sex determination, known as *haplodiploidy*. Female offspring develop from fertilized eggs and therefore have two sets of chromosomes (they are *diploid*), whereas male offspring develop from unfertilized eggs and therefore contain only a single set of chromosomes (they are *haploid*) and inherit all of their genes from their mothers. Daughters will receive identical sets of genes from their father, as he only has one set to give. The other half of the daughters' genes come from the mother, so a daughter will have a 50% chance of sharing each of her mother's genes with a sister. Overall, provided they have the same father, sisters will therefore share around 75% of their genes. In contrast, brothers have a degree of relatedness of only 1/2 with their siblings, as they will have a 50% chance of sharing a particular gene derived from their mother. Sisters are therefore more closely related to each other than in animals lacking this form of sex determination, and any genes underpinning altruistic behaviour are more likely to be passed on by sisters helping to raise female siblings (with a degree of relatedness of 3/4) than by sisters raising their own offspring (with a degree of relatedness of 1/2). In Hymenoptera colonies, the workers that devote themselves to foraging, nest building, defence, or brood-rearing are generally all females.

Hamilton coined the term *inclusive fitness* to capture the idea that the reproductive success of an individual depends not only on how many offspring it has, but also on the extra fitness it can gain through assisting relatives to reproduce. Later, the term *kin selection* came into use to describe selection that takes account of benefits to other relatives as well as immediate descendants. Kin selection is not confined to the Hymenoptera and can be generally applied to any situation in which an individual behaves in an apparently altruistic way towards closely related kin to enhance their inclusive fitness. Subsequently, George Price, a brilliant self-taught maverick, showed that kin selection could be regarded as a special case of group selection. Price had carefully studied the population genetic models in Hamilton's papers on the evolution of altruism, as he was concerned about the implications for humanity if nepotism turned out to be a fundamental human trait (Segerstrale, 2013). In attempting to prove that Hamilton must be wrong, Price (1970) derived his own mathematical approach that could be applied to a much broader set of evolutionary scenarios, including ones in which spiteful behaviour could emerge. Hamilton immediately embraced Price's work and the idea that natural selection can act simultaneously on multiple levels. Based on Price's equations, Hamilton showed that, if altruists favour each other, groups of altruists could be more successful than groups of individuals that did not engage in altruism. The idea that selection could act on groups of individuals was subsequently developed into the theory of multi-level selection (Sober and Wilson, 1998).

Robert Trivers was to describe kin selection as 'the most important advance in evolutionary theory since Darwin' (1985, p. 47), and Wilson regarded it as 'the most important idea of all' (1994, p. 315). As one of the world's greatest experts on the social insects, Wilson was among the first to realize the significance of Hamilton's research. He became an enthusiastic champion of Hamilton's work and, together with Dawkins, can take credit for bringing the idea of kin selection to prominence. In *Sociobiology*, Wilson (1975a) showed how kin selection could be used to explain why many primates, social carnivores, cooperatively breeding birds, and aphids also help their mothers to raise their sisters, as well as why many mammals and birds place themselves at risk by giving warning calls about predators. Unsurprisingly, when Wilson turned his thoughts to humanity, he attempted to explain why a specific group of human beings should apparently be prepared to forgo direct reproductive opportunities, namely the homosexual community, in terms of kin selection. Developing an idea proposed by Trivers, he speculated that:

> the homosexual members of primitive societies may have functioned as helpers … Freed from the special obligations of parental duties, they could have operated with special efficiency in assisting close relatives. Genes favoring homosexuality could then be sustained at a high equilibrium level by kin selection alone. (1975a, p. 555)

This hypothesis is typical of the final chapter of *Sociobiology*: bold, speculative, and naïvely insensitive to political connotations. Nonetheless, kin selection has since been invoked to explain a great deal of altruistic behaviour, with many successes (Marshall, 2015), although the power of Hamilton's rule to predict animal behaviour, and its ability to be tested, has also come into question (Nowak et al., 2017).

Conflict between parents and offspring

The groundbreaking works of Hamilton were followed by equally influential papers by Robert Trivers at Harvard University. Despite suffering from painful spells of mental ill health (Segerstråle, 2000), Trivers single-handedly devised a wealth of sociobiological theory in a few fertile years in the early

1970s. Like Hamilton, Trivers revealed his academic abilities while still a graduate student, in his case in the same Harvard department as Wilson. He was described by Wilson as an individual with dazzling intellect who periodically would burst into his office and let loose a flood of ideas, some wild and some brilliant. Wilson likened a conversation with Trivers to 'taking a mind-altering and possibly dangerous drug' (1994, p. 325) and confessed that two or three hours with Trivers left him exhausted for the day. Trivers's extraordinary contribution stemmed from his mastery of the gene's-eye view perspective.

Trivers contended that differences in degrees of relatedness will result in conflicts of interest between individuals. In two pioneering papers, 'Parental investment and sexual selection' (1972) and 'Parent–offspring conflict' (1974), Trivers reasoned that, in diploid species, parents should favour equal investment in all of their offspring if the costs of production are equal, whereas offspring should favour increased investment in themselves rather than in current or future siblings. The logic behind this idea is that parents are equally related to all of their offspring, whereas offspring have greater interest in themselves than in their siblings. Trivers suggested that natural selection will favour traits in offspring that helped them to get as much food and support as possible from their parents before being forced to become self-sufficient. In contrast, selection will favour parental behaviour that strikes a balance between their investment in current offspring and saving some energy and resources for the next litter, clutch, or child.

Trivers noted how, in many birds, the chicks will vociferously and energetically beg their parents for food around the time of fledging. He also reported how, in langurs, baboons, and rhesus monkeys, an apparent conflict over weaning often lasts for several weeks. The infant utters a series of piercing cries in its effort to beg for milk from its mother, or hitch a ride on her back, whereas the mother frequently pushes the head away from the nipple or pulls the infant away from her body. Trivers interpreted the 'temper tantrums' of young birds and monkeys as an attempt by offspring to manipulate the parent into prolonging the period of parental investment. Prior views of parent–offspring squabbles had treated them as either a non-adaptive consequence of the rupture of the parent–offspring bond or a device promoting the independence of the shy young animal. In contrast, Trivers interpreted these conflicts as the outcome of natural selection operating in opposite directions on the two generations.

The exact dynamics of parent–offspring conflict have proven less easy to investigate, and some assumptions of the theory, including that parent and offspring behaviours are heritable and have an independent genetic basis, and that parent and offspring behaviours have reached equilibrium, remain open to question (Smiseth et al., 2008). Nonetheless, Trivers's ideas provided huge impetus for further work by biologists and led to a fresh interpretation of parent–offspring interactions in humans. Could the temper tantrums of young children be an attempt to manipulate their mothers into prolonging breastfeeding and other forms of parental investment? It turns out that tantrums in two-year-old children seem to be concerned with the process of establishing autonomy from the parents, rather than with conflicts of interest over weaning (Bateson, 1994). Nonetheless, Trivers's insights stimulated a battery of studies investigating parent–offspring conflict in primates, including humans, and provided the impetus for important advances. For instance, during human pregnancy, the mother and foetus may be in conflict over the optimal levels of resources that are transferred across the placenta as a result of these antagonist selection pressures (Haig, 1993). Such maternal–foetal conflicts could explain the specific patterns of blood vessels in the placenta and some maternal hormone responses that characterize pregnancy, and illnesses such as pre-eclampsia and gestational diabetes could perhaps be described as occurring when the mother is unable to mount a sufficiently strong physiological response to the demands of the foetus. In contrast, for other aspects of placental function, such as the transfer of antibodies, the interests of mothers and offspring are more closely aligned, and conflict is likely to occur.

In 1973, Trivers published a paper, together with Dan Willard, in which they suggested that parents may be selected to invest different amounts of resources in sons compared to daughters, if that would maximize the number of grandchildren that were produced. Data consistent with the Trivers–Willard hypothesis were subsequently reported in several animal species, notably red deer, although the expectations were not confirmed in other species, including non-human primates (e.g. Brown and Silk, 2002). This hypothesis was used by anthropologist Mildred Dickemann (1979) to investigate why some human parents prefer sons or daughters. In many parts of the world, girls are more likely to be killed, abandoned, or deprived of food or medicine than are boys. For example, historically in China, for every 100 daughters reported to have been born, around 114 sons have been registered (compared to 105 in most Westernized countries), with the majority of these missing females having been aborted during pregnancy or killed after birth (Clarke, 2000). Using examples from India, China, and medieval Western Europe, Dickemann suggested that a preference for sons is related to the pattern of wealth inheritance. For example, in early nineteenth-century India, where all individuals were born into a closely defined socioeconomic class (or caste), sons inherited the wealth of the family, whereas daughters were expected to marry into a higher social stratum. The British colonizers of India were puzzled by the lack of daughters in the very highest-ranking families, those of the Rajput subcaste, until they realized that most daughters in this caste were being killed soon after birth. The daughters could not be married and consequently would detract from the abundant wealth that would otherwise be inherited by the sons.

Dickemann found that female infanticide was more common among high- than low-caste families, allowing her to explain why human sex ratios varied with socioeconomic status in these countries by using Trivers and Willard's hypothesis. However, Dickemann's prediction that families of low socioeconomic status would favour daughters over sons was not upheld by the data from nineteenth-century India, possibly because daughters entailed the costs of a dowry, a sum of money, or other currency that is required in order to be accepted as a wife. Indeed, testing predictions about sex-biased parental investment has proven to be more complicated than first expected in humans and other animals (Brown, 2001). In addition, sex ratios at birth in human beings have been linked with numerous factors, including parental age, extreme fluctuations in environmental temperature, and the stress of natural and human-related disasters (Navara, 2018), yet the underlying physiological mechanisms remain speculative and unconfirmed. Nonetheless, the application of Trivers and Willard's idea to human beings provided a stimulating new perspective on the question of why parents may treat sons and daughters differently.

Reciprocal altruism

In an article published in 1971, Trivers introduced the idea of *reciprocal altruism* to explain cases where unrelated individuals cooperate with each other. He suggested that, if unrelated individuals interact over an extended period of time, an altruistic behaviour that was initially costly to the actor but beneficial to the recipient could be selected, if there was a high probability that the altruistic act would be reciprocated between the two individuals on a future occasion. Over time, both individuals would gain more than if they had not cooperated at all. However, the difficulty that must be overcome is the tendency for individuals to cheat and not to reciprocate. Reciprocal altruism may therefore be predicted to occur more frequently in cases where individuals interact with each other on a regular basis and maintain a memory of previous interactions, so that cheating individuals would be unlikely to receive altruistic benefits in the future. Nonetheless, more subtle forms of cheating, such as never reciprocating quite as much as one receives, might be expected, if individuals can get away with it.

One of the best-known studies of reciprocal altruism in nature was carried out on vampire bats by Gerald Wilkinson (1984) of the University of Maryland. After a night of foraging, some individual vampire bats returned to their hollow tree roosts hungry and low in body weight, having failed to find a source of blood. Wilkinson observed other group members in the tree regurgitating food to these individuals, who otherwise risked starvation. As the bats were found to live in relatively stable groups, and returned to the same roost site each morning, this food exchange may have involved an element of reciprocity. Here, the cost of blood donation is small; however, it can make the difference between life and death to the recipient, who is sure to get the opportunity to return the kindness. While similar evidence for reciprocal altruism has been documented in fish, birds, monkeys, and apes, such data are often not conclusive, and, in most reported cases, alternative explanations, such as mutualisms (in which both individuals obtain immediate benefits), manipulations (in which one individual is forced to provide assistance to another) or kin selection, have often sufficed (Clutton-Brock, 2009).

In fact, human beings may be the animal in which reciprocal altruism most commonly occurs. Trivers (1971, 1985) argued that reciprocal altruism is likely to have evolved in the small stable social groups inhabited by human ancestors over the last few million years. The system that evolved should allow humans not only to reap the benefits of altruistic exchanges, to protect themselves from gross and subtle forms of cheating, but also to practise forms of cheating where profitable. Moreover, he suggested that selection for reciprocal altruism provides an explanation for certain characteristics of humans. For instance, the need for *friendship* is argued to be adaptive because it motivates us to find and associate with individuals with whom we can trade altruistic acts. *Moralistic aggression*, on the other hand, has potentially evolved so that cheaters will not go unpunished, whereas *gratitude* on the part of the recipient of a kindness is perhaps adaptive because it makes the donor believe that the beneficiary is likely to reciprocate on a future occasion. Finally, in complex social systems that practise reciprocal exchange, a *sense of justice* could be used to judge the behaviour of others. Although it is possible to construct alternative explanations for these traits in humans, Trivers' explanations were both intuitive and compelling. As a result of Trivers' work, economists have become particularly interested in whether humans act reciprocally when they bargain over the distribution of money or resources.

Evolutionary game theory

In turn, ideas from the field of economics were influential in the development of sociobiology. *Evolutionary game theory* is a way of thinking about evolution when the advantage of behaving in a particular manner depends upon what other individuals are doing. Building on earlier game theory ideas, Maynard Smith and Price (1973) pioneered this evolutionary approach. The goal of the exercise is to try to work out which behaviour is the most stable strategy, on the assumption that over millions of years of evolution this is what would have evolved. For instance, in deciding whether to engage in a fight over a resource, an individual may adopt the strategy of 'always attack', 'never start a fight', or perhaps 'always attack when challenged'. Other strategies are conditional, for example, 'attack only if the opponent is smaller' or 'retreat only if the opponent is larger'. If all of the possible strategies are pitted against each other, for example, on a computer or by constructing a mathematical model, the winning strategy can be determined. This strategy is known as the *evolutionarily stable strategy* (ESS), and, if it is adopted by all members of the population, no other strategy could replace it. Alternatively, the ESS may actually be a set of strategies that emerge and are likewise robust to being replaced by another set of strategies.

Evolutionary game theory was originally applied to the study of animal conflicts over resources, notably by Geoffrey Parker. As a graduate student at the University of Bristol, Parker (1970) studied *Scatophaga* dung flies, which lay their eggs in animal dung, often a cowpat, where the larvae feed on the larvae of other insects. Male dung flies wait at the cowpat for a female dung fly to arrive, and females then mate with multiple partners. The shape of the female's sperm storage organs ensures that the sperm from the last mating partner are most likely to fertilize the eggs. In order to ensure his reproductive success, a male fly could wait with the female until she lays the batch of eggs, but then he must fight off other males who attempt to displace him. So, the dilemma for the male fly is how long to guard a female with whom he has mated, and when to move to another cowpat in search of alternative matings. This situation resembles a scenario, or 'game', known as the 'war of attrition', in which individuals trade off the strategic gains from outlasting competitors against other gains they could have acquired had they not waited. Parker was able to use mathematical models to predict the duration of contests between males, and the probability that an individual would give up and leave the patch (Parker and Thompson, 1980). Parker's dung fly data were later used by Maynard Smith (1982) to test the predictions of other evolutionary game theory analyses.

Evolutionary game theory has subsequently been used to investigate how individuals might behave in a wide variety of situations, including whether to forage or steal food, when to cooperate with another individual, and what information to share with others. In each case, the conclusion that a particular ESS should be reached can be tested by investigating whether something like the ESS is observed in natural populations of animals, as in the case of the dung flies. The quantitative rigour that the ESS framework imposes on thinking in animal behaviour has undoubtedly made it an indispensable tool for the study of adaptation.

In *The Selfish Gene*, Dawkins (1976) made ESSs a cornerstone of some of his argument. For instance, he used evolutionary game theory to illustrate the conflicts between the sexes that may arise over the amount of parental investment that males and females are willing to provide to offspring. Developing ideas introduced by Trivers in 1972, Dawkins pointed out that, although both parents may want offspring, for each there may be some advantage to investing less than their fair share of time and resources in caring for these offspring. This prediction is based on the logic that, if they can manipulate their partner into bearing the bulk of the costs of a successful rearing, a parent has still effectively passed on their genes but has extra time and resources to devote to further reproduction. Trivers had pointed out that, when species are classified according to the relative parental investment of the sexes in their young, the male's only contribution is his sex cells in many vertebrate species, while the female's contribution can exceed the contribution from the male by a large amount.

Dawkins suggested that one strategy that a female might adopt to get round this problem (labelled 'coy') is to seek out reliable males and subject them to an extended courtship to assess their fidelity. In fact, Dawkins proposes two female strategies ('coy' and 'fast') and two male strategies ('faithful' and 'philanderer'). Coy females will not copulate until after a long courtship, whereas fast females will copulate immediately. Faithful males are prepared to undergo the extended courtship and will help rear the young. In contrast, philanderers are not interested in courtships and will leave if they do not copulate immediately, without helping to raise any young. For illustrative purposes, Dawkins allocates arbitrary values for the various costs and benefits: 15 units for each offspring successfully raised, 20 units for the cost of rearing, and 3 units for the cost of courtship. With these particular values, a mixed ESS is reached in which five out of six females are coy and five out of eight males are faithful, or when each female is coy five-sixths of the time and each male faithful five-eighths of the time. Thus, in this particular instance, males are more likely to be promiscuous than females are to be fast.

At the end of this discussion, Dawkins reflects on the extent to which this reasoning applies to human beings. Although the values for the costs and benefits are arbitrary, and different values would yield different results, and despite acknowledging that promiscuity is probably more affected by culture than by evolved dispositions in humans, Dawkins concluded 'it is still possible that human males in general have a tendency towards promiscuity' (1976, p. 164). Evolutionary game theory has been less commonly applied to life history strategies of human beings than those of other animals. However, this methodology still remains important for those interested in how humans decide between choices when the outcomes of those choices are influenced by the decisions of others (Gintis, 2009; Schecter and Gintis, 2016), and it has been instrumental in the creation of the new field of evolutionary economics.

The human sociobiology debate

The revolutionary ideas of Williams, Hamilton, Trivers, and Maynard Smith were of huge importance for the study of animal behaviour. The controversial aspect was the application of sociobiology to human beings, and Wilson, in particular, received much of the attention in this regard. Wilson later summarized his ideas on human sociobiology as follows:

> Human beings inherit a propensity to acquire behavior and social structures, a propensity that is shared by enough people to be called human nature. The defining traits include division of labor between the sexes, bonding between kin, incest avoidance, other forms of ethical behavior, suspicion of strangers, tribalism, dominance orders within groups, male dominance over-all, and territorial aggression over limiting resources. Although people have free will and the choice to turn in many directions, the channels of their psychological development are nevertheless ... cut more deeply by the genes in certain directions than in others. While cultures vary greatly, they inevitably converge toward these traits. (1994, pp. 332–333)

These views provoked strong opposition, and before long the dispute had become a media event, particularly in North America. Almost immediately, a vocal countermovement of hostile critics of human sociobiology arose. Anthropologists, psychologists, sociologists, and some prominent biologists bitterly repudiated the sociobiologists' findings, lambasted their methods, and charged them with prejudicial storytelling. The history of this controversy has itself resulted in numerous articles and books (e.g. Segerstråle, 1986, 2000). Here, we will provide only a short account of the political opposition to human sociobiology and spend more time discussing the scientific arguments. We reserve a critical evaluation of Wilson's conception of 'human nature' for Chapter 8.

The immediate opposition that arose against the ideas in the final chapter of *Sociobiology* were from politically active academics. A Boston-based collective of scientists and social scientists came together to form the Sociobiology Study Group, which soon affiliated itself with Science for the People, a nationwide organization of activists begun in the 1960s to expose the misdeeds of scientists. At a time when student demonstrations against the Vietnam War were commonplace and following on from a recent melee over race and intelligence testing, motivating students to protest against allegedly subversive or dangerous scientists was not difficult. The Sociobiology Study Group was dominated by Marxist and left-wing scholars from Harvard University. Two of the most prominent and vocal were evolutionary biologists Richard Lewontin and Stephen Jay Gould, both of whom worked in the same building at Harvard as Wilson. In fact, the group of scientists and social scientists most openly critical of human sociobiology met together in Lewontin's office, directly below Wilson's.

In a letter published in the *New York Review of Books* on 13 November 1975, the Sociobiology Study Group declared that human sociobiology was not only unsupported but tended to provide a genetic justification of the status quo and perpetuated inequalities on the basis of sex, class, and race. The group accused sociobiology of being 'reductionist', 'biologically determinist', and motivated by ignorance and chauvinism, took issue with the hypothesis that society reflected biological imperatives, and linked sociobiology to disturbing former applications of evolutionary theories:

> These theories provided an important basis for the enactment of sterilization laws and restrictive immigration laws by the United States between 1910 and 1930 and also for the eugenics policies which led to the establishment of gas chambers in Nazi Germany. The latest attempt to reinvigorate these tired theories comes with the alleged creation of a new discipline, sociobiology. (Allen et al., 1975)

Wilson clearly believed that good scientists should have the courage to address difficult issues and to keep going regardless of any resistance. Convinced that what he had said in *Sociobiology* was justifiable, Wilson went on the offensive (e.g. Wilson, 1976). He castigated his critics as political extremists who perpetuated the myth of the mind being a blank slate at birth (*tabula rasa*) only because it was consistent with their naïve dream of a perfect society. At the same time, he extended his research into human behaviour in his next book, *On Human Nature* (1978), which was an immediate bestseller and won a prestigious Pulitzer Prize. Wilson continued to make bold claims that were to keep him at the heart of the controversy, suggesting, for example, that differences in the behaviour of men and women reflected past evolutionary events and could only be eradicated at some cost to society. The debate became highly charged and politicized, and feelings ran so high that, at a major scientific meeting in 1978, a group of demonstrators took over the stage when Wilson was about to speak, shouting out that Wilson was a racist, and then dumped a pitcher of iced water on his head.

One critic whom Wilson could not dismiss as biologically naïve was his colleague, Richard Lewontin. The conflict between these two members of the same Harvard department, each an outstanding scientist and an authority on evolutionary biology, born in the same year, is one of the most beguiling features of the sociobiology controversy. Ironically, it was largely through Wilson's efforts that Lewontin was recruited to Harvard in the early 1970s, only for Lewontin seemingly to bite the hand that had fed him. However, Wilson later acknowledged Lewontin as a worthy adversary and suggested that, without Lewontin, the controversy would not have attracted so much attention (Wilson, 1994). Lewontin gave countless public lectures criticizing sociobiology and wrote endless hostile reviews of sociobiological books (e.g. Lewontin, 1977). There was no doubting Lewontin's credentials; he was one of the world's leading population geneticists, who used molecular biology techniques to answer evolutionary questions, and Gould once described him as the cleverest scientist he had ever known. Lewontin had always been very open about his strong political views and willing to stand by his principles. Lewontin had been elected to the National Academy of Sciences at a relatively young age, only to resign in protest over the Academy's sponsorship of military research.

In the cauldron atmosphere of the sociobiology debate, it was easy to take sides and dismiss Wilson as motivated by prejudice or Lewontin by his Marxist ideology. In reality, their differences were not solely related to politics but also concerned how science should be done (Ruse, 1999; Segerstråle, 2000). Wilson was the kind of scientist who relished the challenge of major problems, saw the big picture, and constantly wanted to push fields forward by developing and synthesizing new theory. In contrast, Lewontin was much more cautious, suspicious of sweeping statements and unsupported speculation, and deeply sensitive to how vulnerable biological arguments are to abuse.

For Lewontin, science had to be as correct as possible, because mistaken scientific theories lent themselves to political abuse, a belief which, as we saw in Chapter 2, can find justification in the history of using evolutionary theory to interpret humanity. As the sociobiology debate raged, biologists also argued over whether complex biological phenomena could be understood at the level of molecules, and, here too, Wilson and Lewontin found themselves on opposite sides of this debate. Human sociobiology was charged with extrapolating recklessly from genes to behaviour without any clear understanding of the potential molecular, cellular, physiological, and social mechanisms involved.

Among many social scientists, one seemingly insurmountable problem for human sociobiology was the special status that was deemed to characterize the human species, based largely on our capacity for culture and language. To Wilson's credit, he was prepared to take on board some of this criticism. In his autobiography, he stated that 'it was obvious to me that human sociobiology would remain in trouble, both intellectually and politically, until it incorporated culture into its analyses' (1994, p. 350). In 1979, Wilson was joined by a postdoctoral researcher called Charles Lumsden, a theoretical physicist, and they jointly developed mathematical models exploring the relationship between genes and culture. Within two years, Lumsden and Wilson published *Genes, Mind, and Culture*, which acknowledged that human culture has socially transmitted features that make it distinct from other aspects of the human phenotype. Human behaviour was described as being influenced by elements of culture, called 'culturgens', that are transmitted between individuals (e.g. particular ideas, beliefs, or behaviour patterns). Whether an individual adopted a particular culturgen was modelled as depending upon the characteristics of that individual's brain, which was framed as subject to genetic biases during development. Thus, even though individuals learn aspects of their culture, humans were described as being programmed to acquire some culturgens more easily than others, as a result of natural selection favouring individuals with rules that bias them towards adaptive behaviour.

In their book, Lumsden and Wilson (1981) also argued that culture can affect the rate of genetic evolution, which was consistent with mathematical models being generated by other researchers. Wilson (1994) confesses to being puzzled by the fact that this work was largely ignored by academic colleagues. However, the book was published in the midst of the sociobiology debate and was never likely to be embraced or judged objectively. When one also considers its highly technical nature and the fact that Lumsden's mathematical methods would have been unfamiliar to virtually all readers, *Genes, Mind, and Culture* never stood a chance. For almost everyone, this theory was completely opaque and, as a consequence, their assessment of it was heavily influenced by hostile reviews (e.g. Kitcher, 1985; Maynard Smith and Warren, 1982). Regardless of the reception to this work, Wilson had made a genuine attempt to respond positively to his critics. Unfortunately, by then, it was too late; sociobiology had become a dirty word to many social scientists, and most of them were highly suspicious of Wilson. He had failed to bring to fruition his vision of a new synthesis between biology and the human sciences. For all his credibility in biological circles, Wilson's human sociobiology had been resoundingly rejected by social scientists, the very people Wilson believed could most benefit from it the most.

Critical evaluation

Despite its turbulent beginnings, has sociobiology increased our understanding of social behaviour, or were social scientists right to dismiss its methodologies and ideas? Perhaps the first point to make here is that the ideas provided by Hamilton, Trivers, Williams, and Maynard Smith, which were presented by Wilson in the first twenty-six out of twenty-seven chapters of *Sociobiology* and by Dawkins in *The Selfish Gene*, have revolutionized the study of animal social behaviour by helping to dismiss naïve group selectionism and providing explanations for the behaviour of diverse animal species. The work carried out using these ideas continues under the name of *behavioural ecology* and has blossomed into a highly productive and rigorous field of enquiry (Danchin et al., 2008; Davies et al., 2012). Virtually all of the controversy has surrounded the application of these ideas to human behaviour. We will review the charges of genetic determinism and reductionism, and prejudice and storytelling, and then return to consider the rejection of sociobiology by the social scientists.

Genetic determinism and reductionism

Human sociobiology was most frequently denounced for the sins of genetic determinism and reductionism. In other words, critics angrily chastised sociobiologists for suggesting that the behaviour of individual human beings is determined by genes and that complex behaviour could be reduced to genetic effects. Again, Lewontin was a strong opponent to this view of behavioural development and to the idea of reductionism. For instance, in *Not In Our Genes* (1984), Steven Rose, Lewontin, and Leon Kamin wrote:

> Sociobiology is a reductionist, biological determinist explanation of human existence. Its adherents claim ... that the details of present and past social arrangements are the inevitable manifestation of the specific action of genes. (p. 236)

At first sight this criticism appears unfounded. Wilson had been keen to point out that he was not attempting to show that human behaviour patterns were solely influenced by genes. The behaviour of all animals is a product of the interaction between genes and environment, while learning and experience allow individuals to acquire novel attributes. In other words, genes exert a diffuse influence on human activities, and the arguments over genetic determinism frequently boil down to just how much influence different people think genes have. All of the major protagonists of sociobiology have bent over backwards to be clear that they do not believe in genetic determinism. Yet much confusion still occurs today when the term a 'gene for' a behaviour is used and when the reproductive success of individuals is said to be measured in terms of the number of genes passed on. The possibility that there may be a genetic influence on behaviour, which may be inherited by the next generation, does not necessarily imply that such behaviour patterns are solely determined by one or more genes, or that such behaviour patterns are fixed and inevitable. Rather, this explanation is based on the narrow assumption that, while a multitude of non-genetic factors influence behavioural development, such influences are typically not heritable and hence can be ignored in evolutionary analyses. Nonetheless, many biologists have expressed concerns that the assumption of genetic control and the 'gene for' language might lead to a neglect, or trivialization, of developmental processes and that sociobiological arguments place too much emphasis on the role of genes (see Chapter 8).

In reality, much of the debate was not about genetic determinism at all, but rather about genetic constraints and propensities. Wilson suggested that 'Rather than specify a single trait, human genes prescribe the *capacity* to develop a certain array of traits' (1978, p. 56, italics in original). Whether an individual expresses a particular behaviour pattern may depend upon a myriad of factors encountered by that individual over a lifetime, including social and cultural influences. In contrast to Lewontin, Wilson believed that a reductionist approach to the study of behaviour was appropriate. One dictionary definition of reductionism is 'the belief that complex data and phenomena can be explained in terms of something simpler'. This method of understanding the world is applied throughout science and is often seen more as a virtue than a sin. Wilson acknowledged that seeking out simple, universal rules is the way that he thinks and sees nothing wrong with this approach.

Perhaps what the critics objected to was not reductionism per se, but *inappropriate* reductionism, in which an account of phenomena is given at completely the wrong level. An example would be a theory of social behaviour based at the level of the atom. Some social scientists feel that a genetic account of human altruism, for example, is inappropriate, as in their view genes explain little of the variation in human altruistic behaviour, whereas cultural factors explain much more. However, a counterargument can be provided even for this variant of the reductionism charge. The scientific method is inherently self-policing against inappropriate reductionism, as hypotheses that provide explanations at the wrong level lack utility and are soon replaced by more powerful hypotheses. Nonetheless, this self-policing is never perfect. One example of where reduction can be applied inappropriately is to assume that complex, large-scale phenomena be explained as the sum of their parts. Lewontin (1991) was particularly critical of the neglect of emergent properties:

> By reductionism, we mean the belief that the world is broken up into tiny bits and pieces, each of which has its own properties and which combine together to make larger things. The individual makes society, for example, and society is nothing but the manifestation of the properties of individual human beings. (p. 107)

Lewontin argued that there are properties of society and of social institutions that do not reduce to properties of people and that, sometimes, these levels of complexity and structure cannot be ignored. In a sense, Wilson acknowledges this fact in *Genes, Mind, and Culture* with his treatment of culture as a dynamic process in its own right. However, Wilson clearly regarded culture as being constrained by our biological heritage, as exemplified by his famous phrase that 'the genes hold culture on a leash' (1978, p. 172). The majority of social scientists are still content to assume that culture is responsible for most human behaviour and that the role of genes is small enough to be of little relevance to the study of humans. There is nothing reductionist or deterministic about the challenging of this assumption, and Wilson was surely justified in asking whether there might be adaptive genetic influences that prevail despite the influences of culture.

Prejudice

Some of the statements Wilson made in *Sociobiology* left him wide open to accusations of prejudice in terms of sex, class, and race. Critics and historians alike have concluded that Wilson's writings on humanity reflect the values of his upbringing in the American South and his undergraduate years at the racially segregated University of Alabama. What is apparent is that, in *Sociobiology*, Wilson does make some injudicious statements and that these provoked much hostility. For instance, his views on sex differences often appeared to promote the status quo, and his views on possible differences in mental aptitudes between races were easy targets for attack. Another example is his discussion of whether genes that affect success and an upward shift in status 'would be rapidly concentrated in the uppermost socioeconomic classes' (1975a, p. 554). In fact, the next page reveals that Wilson is quite aware of some of the counterarguments; for example, that society is too fluid and gene flow too extensive to maintain such genetic differences between social class. Yet, while playing the intellectual game of devil's advocate, Wilson seemed prepared to entertain perspectives that would be anathema to more politically astute scientists and had apparently spared little thought for the repercussions of his deliberations.

Wilson claimed to have been ignorant of the possibility of outrage at his work. Maynard Smith stated he had disliked the last chapter of *Sociobiology* and later remarked 'it was absolutely obvious to me—I cannot believe Wilson didn't know—that this was going to provoke great hostility' (Segerstråle, 2000, p. 241). However, in his autobiography, Wilson (1994) reiterated his political naïvety, but did recognize that:

> Mine was an exceptionally strong hereditarian position for the 1970s. It helped to revive the long-standing nature–nurture debate at a time when nurture had seemingly won. The social sciences were being built upon that victory. (p. 333)

Many critics of sociobiology were concerned that, irrespective of whether sociobiologists themselves harboured prejudices, sociobiological arguments were vulnerable to racist and prejudicial interpretations. When, in 1981, British biologist Steven Rose wrote a letter to the journal *Nature* revealing that an extreme right-wing organization had been using sociobiological writings to support their racist creed, for many these fears appeared justified. Rose challenged leading sociobiologists to disassociate themselves from these neo-Nazi views, and Maynard Smith, Dawkins, and Wilson all immediately wrote firm replies stating that there could be no justification for racism in sociobiology.

The sociobiological claim that human social organization reflects a history of natural selection led some advocates and critics alike to the conclusion that the current state of society is in some sense optimal. Wilson also warned that humans can only manufacture a society based on full equality at a cost and asserted that our genetic heritage may render it impossible to mould society in certain directions. For those who regarded American society in the 1970s as riddled with race, class, and gender prejudices, this message was highly unsavoury and appeared to be defeatist. However, Wilson's view was not shared by all human sociobiologists. In *Darwinism and Human Affairs* (1979), biologist Richard Alexander of the University of Michigan stated that an evolutionary interpretation of human history does not imply a deterministic future and that conscious awareness of our biology allows us to release the bind to our history of fitness maximization. Moreover, Dawkins (1981), in his response to the neo-Nazi article exposed by Rose, stressed that genetically inherited traits were far from unmodifiable:

> What is really wrong with the National Front quotation is not the suggestion that natural selection favoured the evolution of a tendency to be selfish and even racist. What I object to is the suggestion that if such tendencies had evolved they would be *inevitable* and *ineradicable*; the suggestion that we are stuck with our biological nature and can't change it. (p. 528, italics added)

Dawkins went on to charge the critics of sociobiology with propagating the 'myth' that sociobiologists believed in the inevitability of genetic effects.

Storytelling

Perhaps the most telling criticism of sociobiology is that many of the hypotheses were no more than plausible stories about the origin of human behavioural traits. Rose, Lewontin, and Kamin (1984) complained that:

> Imaginative stories have been told for ethics, religion, male domination, aggression, artistic ability, etc. All one need do is predicate a genetically determined contrast in the past and then use some imagination, in a Darwinian version of Kipling's *Just So Stories*. (p. 258)

Ironically, the problem stems from the fertile nature of evolutionary reasoning. Inventing evolutionary stories is a seductively easy exercise. If we were to attempt to explain human sex differences in average height, for example, we could come up with numerous evolutionary hypotheses. For instance, men may be taller on average than women because, in the past, females preferred to mate with tall males, or perhaps extra height gave tall males an advantage while hunting in the savannah, searching for prey, or throwing spears, or gave some advantage during fights with other males. Perhaps extra shortness gave females an advantage while gathering, making it easier to collect plant material on the ground, or leaving them less visible to predators. Height has the advantage that it is a characteristic that is manifest in the fossil record, that it is easily quantified in contemporary populations, and that the sex difference is unlikely to result from social or economic variation. Behavioural and psychological attributes such as promiscuity or intelligence are not even subject to these constraints on imaginative reasoning. Developing hypotheses is a fundamental part of the scientific process, and part of the enduring appeal of evolutionary theory is that it is such an effective instrument for doing so. However, the opportunity for creativity only makes it even more important that evolutionary hypotheses should not only be potentially testable but actually tested.

Let us return to Wilson's kin selection explanation for homosexuality. Rose and colleagues (1984) pointed out the weaknesses in this argument. First, there was no evidence that in past environments homosexuals had fewer offspring than heterosexuals. Second, there was no satisfactory evidence that homosexuality had a genetic basis. Third, there was little evidence that homosexuals, either now or in our evolutionary past, helped their relatives to raise offspring any more than did heterosexuals. In the intervening years, evidence for biological influences on homosexuality has accumulated, but it remains contentious. There is also only limited evidence that homosexuals have fewer children on average than heterosexuals (Kirkpatrick, 2000), and, in most societies that have been studied, homosexual men are reportedly no more likely than heterosexual men to channel extra resources to family members (Bobrow and Bailey, 2001). Given the politically sensitive nature of the topic, it is easy to see how the superficiality of Wilson's explanation could be regarded as irresponsible.

The direct comparison between the behaviour of human beings and other animals had been a method commonly used by the popularizers of human ethology. Wilson had tried to introduce a little rigour into the comparative analysis of human behaviour. He clearly believed that he could do better than predecessors such as Desmond Morris and Robert Ardrey, whose 'particular handling of the problem tended to be inefficient and misleading' (Wilson, 1975a, p. 551). Wilson suggested that traits that vary from species to species, or genus to genus, are so evolutionarily unstable that it is foolhardy to use them to make comparative inferences about humans. Only those characters that are constant at the taxonomic level of family, or order, may be sufficiently stable to have persisted in relatively unaltered form during the evolution of modern humans. Such conserved traits, Wilson suggested, might warrant an evolutionary explanation. Although this careful approach is admirable, much of the comparative work in human sociobiology was based on views of primate social behaviour and hunter–gatherer lifestyles that are now interpreted differently. That is not to say that comparative methods, which have since come a long way (Nunn, 2011), cannot be employed to draw testable inferences about human behaviour. Even so, because all of our close relatives are extinct and because we know so little about other species in the *Homo* genus or other hominins, it is difficult to be sure that behavioural traits that appear to be homologous with apes (i.e. inherited from common ancestors) really are so.

The work of Sarah Blaffer Hrdy provides examples of how taking a careful comparative approach can help us to understand our own species and dispel outdated views of human behaviour. Hrdy was an anthropology student at Harvard during the time of the sociobiology debate, and her mentors included Wilson and Trivers, as well as the biological anthropologist Irven DeVore. During an undergraduate lecture attended by Hrdy, DeVore mentioned a report by Japanese primatologists working in India that described adult male langur monkeys grabbing infants from their mothers and biting them to death. This behaviour was assumed to be 'pathological' and caused by high population densities. Hrdy was intrigued and carried out her postgraduate work on these langurs. Her studies in the field showed that male attacks on infants only occurred when new males entered the breeding group. Using Trivers' framework, Hrdy (1977) suggested that males would be selected to eliminate unweaned infants that were not their own, as females would then ovulate sooner than if the infant had lived and continued to suckle. The apparent willingness of female langur monkeys to mate with infanticidal males was viewed by Hrdy as an adaptive strategy on the part of females in response to the high turnover of adult males in the group. The mother of an infant would make attempts to prevent the infanticidal attack yet would often mate with the male that had recently killed her offspring.

In 1981, Hrdy published *The Woman that Never Evolved*, in which she put forward her views on the evolution of women and other female primates. These ideas followed a tradition of other women, including Antoinette Brown Blackwell (1875) and Clémence Royer (1870), who had responded to the publication of Darwin's *The Origin of the Species* by pointing out his lack of emphasis on the evolution of female behaviour. However, not until the end of the twentieth century was the behaviour of female primates to receive the attention that it deserved. In her book, Hrdy pointed out that the view of women and other female primates as sexually and socially passive animals was just not backed up by the available evidence. Hrdy demonstrated how female primates have strategies of their own and showed that the relationships between female primates greatly influence the dynamics of social groups. In a later book, *Mother Nature: Natural Selection and the Female of the Species* (1999), Hrdy argued that infanticide by males may be less common in humans than in other primates, such as langurs, gorillas, and chimpanzees, but that the potential threat from others may partially explain why human infants show a fear of strangers. Hrdy's work shows how the study of animal behaviour can increase our understanding of human behaviour and highlights the fact that sociobiology can lead to the dispelling, rather than the bolstering, of biased views of human evolution.

The rejection of sociobiology by social scientists

For most social scientists, the real problem with sociobiology was not that it was reductionist, or deterministic, or prejudicial, nor that sociobiologists were encroaching on their territory. Rather, it was that too much human sociobiology was dilettante. In their enthusiasm, human sociobiologists capriciously flitted from one topic to the next, often concocting superficial stories without ever stopping to develop a solid understanding of the topic, read the social science literature, or consider alternative non-evolutionary explanations. In fact, the attacks on their work from the social scientists may even have caused the early human sociobiologists to close ranks and to avoid criticizing each other's work. Had sociobiologists been more responsible in their application of evolutionary theories of human behaviour (for instance, by asking whether they had evidence for their suppositions, considering the merits of non-evolutionary explanations, and using the data and insights collected by social scientists), they may have been less likely to provoke a negative reaction.

The tragedy is that many of the good ideas generated by sociobiology were dismissed because of its failings. In his autobiography, Wilson complained that the sociological or cultural models of his critics were assumed to be true unless proven false beyond any possible doubt, whereas sociobiological hypotheses were assumed to be false unless the evidence was completely unassailable in their support. Although Wilson has a case, there is a strong argument that this bias is justified. Perhaps human behavioural scientists should have as a starting point, or null hypothesis, the assumption that all types of society are possible and that no behavioural differences between subsections of the population are impossible or onerous to eradicate. If a higher standard of science were maintained by those taking an evolutionary approach to human behaviour, evolutionary explanations might be less open to abuse.

Sociobiology has undoubtedly contributed to an understanding of human behaviour, particularly with regard to topics such as cooperation, conflicts of interest, and parental investment. Moreover, sociobiology provided a new set of methods with which to explore human behaviour, including the gene's-eye view, kin selection, reciprocal altruism, and evolutionary game theory. The 'selfish gene' perspective was a major advance in thinking about animal behaviour and evolution, and it applies equally to humans. What should be remembered is that 'selfishness' as a human trait is not implied by the 'selfish gene' view, and cooperative behaviour is equally compatible with sociobiological reasoning. For example, sociobiological theory has shown that altruistic behaviour in humans may be explained in selfish-gene terms. The observation that certain human behaviour patterns may have evolved also does not imply that they are 'right', a mistake that sociobiologists have repeatedly cautioned against. In *The Selfish Gene*, Dawkins wrote, 'I am saying how things have evolved. I am not saying how we humans morally ought to behave' (1976, pp. 2–3), and Wilson stated, 'There is a dangerous trap in sociobiology, one which can be avoided only by constant vigilance. The trap is the naturalistic fallacy of ethics, which uncritically concludes that what is, should be' (1975b, p. 48).

A comparison of human beings with other social animals, particularly other primates, can reveal what is unique to the human species and what is similar to other animals. Moreover, a rigorous analysis of animal behaviour across a broad range of species allows the abstraction of general principles that may also apply to human beings. In addition, the detailed study of animal behaviour and careful use of similar methodologies with human beings may help us to understand whether current views that, for example, particular sex differences in behaviour are deeply rooted in our animal heritage, are based on false or limited understandings of animal social behaviour. Evolutionary biology can result in prejudicial myths being dispelled, as well being used to support them. However, the potential for scientific ideas or findings to be abused by those with particular political leanings or prejudices means that human evolutionary scientists continue to have a responsibility to think carefully about how they describe their findings and the terminology that they deploy. As demonstrated in the historical review in Chapter 2, and in the fallout from the sociobiology debate, ideas derived from this research field are commonly used to define what it means to 'be human', an issue to which we return in Chapter 8.

A new dawn?

In 1977, Wilson received the National Medal of Science from President Carter for his contributions to the new discipline of sociobiology. Wilson was repeatedly nominated for a Nobel Prize for starting the field and his many other contributions to science. Perhaps his unique combination of creativity, courage, and political naïvety will eventually be regarded as the catalyst for his dream of an integration of biological and social sciences. While this vision has yet to be fully realized, there are signs that related approaches, which emanated from the groundbreaking work of the sociobiologists, have greater resonance with social science thinking. These contemporary approaches will be described and evaluated in the remaining chapters.

The first strand to emerge was *human behavioural ecology*, mainly carried out by biological anthropologists who have explored to what extent human behaviour is adaptive, under the assumption that much culture is evoked by various features of the social and ecological environment. Human behavioural ecologists are generally from an anthropological background, rather than being biologists that have fleetingly turned their hands to the human sciences, a shift that in itself has led to significant advances. A few years later, the second strand was to emerge, now named *evolutionary psychology*, which has generally been carried out by academic psychologists searching for the evolved psychological mechanisms that underpin the universal mental and behavioural characteristics of humanity. In the final chapter of *The Selfish Gene*, Dawkins presented the idea of the meme, devised as an alternative to sociobiological explanations of culture, while biologists and anthropologists were devising mathematical models of *cultural evolution*. The field of *gene–culture coevolution* was emerging at around the same time, developing ideas and methods along the lines of Lumsden and Wilson's *Genes, Mind, and Culture*. However, this work is practised by some of the population geneticists and anthropologists that had been among sociobiology's strongest critics.

While human behavioural ecology, evolutionary psychology, cultural evolution, and gene–culture coevolution differ in important respects from sociobiology, and proponents of each would regard these differences as a major advance, all of these modern sub-fields germinated their roots in, and owe a debt to, the sociobiological era. In some ways, human sociobiology could be described as ongoing, although many researchers avoid using this term to describe their work as a result of the controversy that followed Wilson's publications. Other researchers, such as Hrdy, have prided themselves in describing their work as human sociobiology out of respect for the positive contributions that the application of methods from animal behaviour to the study of human beings have provided. That sociobiology could spawn a wealth of new evolutionary approaches to the study of human behaviour is a testament to the rich, fertile, and pluralistic nature of its theory. For human sociobiology, there were to be new dawns.

4
Human behavioural ecology

While the human sociobiology debate was raging, a number of anthropologists and biologists decided to go out and test sociobiological ideas with real data from human populations. They started by asking questions such as '*Do human beings exhibit optimal strategies during foraging?*' and '*Do people alter the number of offspring they raise depending upon their environment?*' Their main premise was that human behavioural strategies are adaptive across a broad range of ecological and social conditions. By collecting detailed, quantitative data on the behaviour, customs, and decisions of people living in a range of different locations and social structures, the researchers could test whether their theoretical ideas and predictions matched real-world examples.

Questions about environmental and ecological influences on human behaviour were already being asked within anthropology. For example, ecological anthropologists had been examining the relationship between human societies and their environments, with humans being viewed as just one part of a larger, mostly static, ecosystem (Moran, 1984). However, these researchers had focused heavily on energy flow, rather than behaviour, and often resorted to group-selectionist arguments. In contrast, early *evolutionary* or *Darwinian anthropologists* were interested in whether individual human beings might be able to alter their behaviour flexibly, depending upon present circumstances, to maximize their own reproductive success (Cronk, 1991). Such research is usually now referred to as *human behavioural ecology*, or more broadly as *human evolutionary ecology*, and incorporates the emerging field of *evolutionary demography*. Human behavioural ecologists are interested in how an individual's behaviour is influenced by the environment in which that individual lives and how the alternative behavioural strategies that people adopt produce cultural variability.

The aim of modern Human Behavioural Ecology (henceforth, this phrase will be capitalized to denote this specific sub-field) is to:

> determine how ecological and social factors affect behavioural variability within and between populations. In one sense its hypotheses are viewed as an alternative

to the more traditional anthropological belief in an unspecified force of 'cultural' determination. In another sense, behavioural ecological anthropology can be seen as adding the study of function to investigations of causation, development and historical constraints that were already well established in the social sciences. (Borgerhoff Mulder, 1991, p. 69)

Early Human Behavioural Ecologists were greatly influenced by the progress being made in the study of animal behaviour through the application of new theories of optimization and life history strategies, which linked the behaviour patterns of individuals to their physical and social environment. These researchers, who sometimes had a background in evolutionary biology or animal behaviour research, began studying humans as if they were any other animal species, observing what individuals did in their real lives and comparing the data to predictions derived from evolutionary hypotheses. As early as 1956, John Haldane had argued that the behavioural differences shown by contrasting human groups were adaptive responses to particular environments and that these different patterns of behaviour were exhibited by human beings with basically similar genetic compositions. This idea was later reiterated by William Irons in the landmark book *Evolutionary Biology and Human Social Behavior: An Anthropological Perspective* (1979), edited by himself and Napoleon Chagnon. At Northwestern University in Illinois, Irons and Chagnon were teaching evolutionary approaches to anthropology students, while, at the University of Utah, Kristen Hawkes, under the guidance of biologist Eric Charnov, also began teaching Human Behavioural Ecology from the late 1970s. The works of zoologists Richard Alexander (1974) and Robert Hinde (1974) provided additional impetus to this emerging field. John Crook, who was a pioneer in the study of animal socioecology and subsequently worked on human behaviour, also had a major impact by showing that social systems could be seen as ecological adaptations (Crook and Gartlan, 1966).

Arguably, the most influential academic in the establishment of this new discipline was Irven DeVore, an anthropologist working in Harvard at the same time as Edward Wilson. DeVore had been a graduate student of the physical anthropologist Sherwood Washburn at the University of Chicago (Kuper, 1994). Although DeVore was a social anthropology student with no previous interest in primates, Washburn sent him to Kenya in 1958 to carry out a pioneering study on the social lives of baboons. Washburn was convinced that studying non-human primates would provide insights into the evolution of human behaviour. He also believed that field studies of contemporary African hunter–gatherers might help researchers to understand the ways in which early humans had adapted to environmental pressures. DeVore and graduate student Richard Lee later initiated an important, long-term study of the !Kung San (or Ju/'hoansi) of the Kalahari. Although Washburn was highly critical of human sociobiology and remained a staunch supporter of group selection, his early mentorship of DeVore played a crucial role in the application of evolutionary theories to human behavioural variation.

At Harvard, DeVore was won over by the ideas of Robert Trivers and became a great supporter during his difficult early career (Segerstråle, 2000). DeVore was also instrumental in promoting sociobiology to a large audience of interested students and academics. Throughout the 1970s, DeVore presided over a thriving research group with a rich atmosphere that resembled an intellectual salon, with distinguished visitors regularly dropping by and discussing sociobiological ideas and methods. For a number of years prior to the publication of Wilson's *Sociobiology* (1975a), DeVore and Trivers had been teaching a course in the anthropology department at Harvard on using evolutionary biology to investigate human behaviour. Although DeVore and Wilson were friendly colleagues, DeVore was not an avid supporter of Wilson's version of sociobiology as applied to humans, partly because he felt that Wilson lacked a comprehensive grasp of the anthropological literature. He later stated:

> When *Sociobiology* was near completion, Ed [Wilson] sent me the last chapter. It was not that I disagreed with him. I wanted him to have written a different book. I felt it was nothing like the whole story. Ed was naïve in many ways in those days. It was not that he had no respect for the social sciences; he had so many other things on his plate. I kept thinking that Trivers and I should have done this! (Quoted in Segerstråle, 2000, p. 81)

From its inception, the goal of Human Behavioural Ecology has been to account for variation in human behaviour by asking whether models of optimality and fitness maximization provide good explanations for the differences found between individuals. Early work, carried out in the late 1970s and 1980s, focused largely on child-rearing strategies and foraging behaviour (Cronk, 1991). For example, researchers asked whether women in hunter–gatherer communities maximize their reproductive success by having relatively prolonged intervals between births (Blurton Jones and Sibley, 1978) and investigated the importance of grandmothers in rearing their grandchildren (Hawkes et al., 1989). Data on diet choices among hunter–gatherer populations were compared to the food items available, and models were produced to test whether individuals were foraging optimally, in terms of gaining the greatest possible number of calories per hour of foraging (Smith, 1985).

Subsequently, the field broadened out to address questions concerned with resource sharing, mating strategies, kinship systems, and divisions of labour (Borgerhoff Mulder, 2024; Gurven, 2004; Koster et al., 2024; Nettle et al., 2013; Winterhalder and Smith, 2000). An overarching premise of the field has been that human beings exhibit an extraordinary behavioural flexibility that allows them to behave in an adaptive manner in a broad range of environments.

Most of the research carried out by Human Behavioural Ecologists has been on small communities in remote regions of the world, often those that have relatively little contact with industrialized societies. The data are collected by direct observation of behaviour and by use of interview and historical material. Thus, the populations being studied, and the methods used, resemble similar research carried out by social anthropologists, particularly ethnographers, but with a different theoretical and epistemological framework. However, such detailed data collection requires substantial time and effort. To complement this research, Human Behavioural Ecologists have made use of existing and historical datasets, such as genealogies from European church records, and have expanded their methods to study people living in industrialized societies (Mattison and Sear, 2016; Nettle et al., 2013; Stulp et al., 2016a). A new field of *applied Human Behavioural Ecology* has also emerged, which investigates how people make decisions about important aspects of their lives, such as health, nutrition, and hygiene, and predicts whether public health interventions are likely to be successful (Gibson and Lawson, 2015). The contemporary field of Human Behavioural Ecology continues to provide a comprehensive framework for studying behavioural variation between individuals, groups, and populations.

Key concepts

Human Behavioural Ecology is characterized by an emphasis on the flexibility of individual behaviour, by the testing of predictions derived from theoretical models, and by an initial assumption that human behaviour consists of adaptive trade-offs or compromises. These ideas are described in the following sections.

Flexibility of individual behaviour

A main premise of Human Behavioural Ecology is that human beings have been selected to optimize their lifetime reproductive success in response to environmental conditions by flexibly altering their behaviour. Researchers in this sub-field assume that many aspects of human behaviour are likely to vary in a facultative manner in response to the particular social and ecological resources to which individuals are exposed. They suggest that a past history of selection will have favoured the ability to adopt the particular strategy that maximizes the difference between the benefits and costs in that particular environment. Such strategies may take the form 'In context X, do *a*; in context Y, switch to *b*' (Smith, 2000). Thus, for Human Behavioural Ecologists, our species is characterized by an extraordinary 'adaptability', a term used by evolutionary biologists to describe the degree to which a species can survive and successfully reproduce in a wide range of environments. This characterization does not equate with the idea that human beings are infinitely flexible, as Human Behavioural Ecologists presume that constraints in underlying genetic, physiological, and neural mechanisms will limit the range of environmental conditions in which adaptive responses can be produced.

The idea of human beings as fitness-maximizing agents does not require that humans make a conscious decision to alter their behaviour in accordance with optimality criteria. Few people consciously calculate how to leave as many descendants as possible. In addition, researchers might not find it practical to measure the actual number of children, or grandchildren, produced by individuals following different strategies because of our long lifespans, which means that more immediate and accessible proxies of reproductive success are often needed. Rather, Human Behavioural Ecologists suggest that individuals may be predisposed to optimize what are assumed to be correlates of lifetime reproductive success, such as the rate of food acquisition or the amount invested in young children, often through subconscious decision-making processes. A key assumption is that a history of selection has endowed our species with a tendency to respond to the environments in which we find ourselves by weighing up the benefits and costs of adopting particular strategies. How specific cues from the environment result in a change in behaviour may depend upon diverse physiological and psychological mechanisms, including the learning of socially transmitted information. However, Human Behavioural Ecologists believe that an understanding of the relative importance of these proximate causal factors is not necessarily required when studying the environmental conditions or fitness outcomes of particular strategies.

Formal models and hypothesis testing

An important aspect of Human Behavioural Ecology is the testing of hypotheses derived from formal or mathematical evolutionary theory. Starting with the premise that behaviour has been selected to optimize reproductive success, models can be produced that predict the optimal pattern of behaviour in a given circumstance. Data are then collected on how human beings really behave, and these data are compared to the predictions derived from the model. Where the data fit the model, the hypothesis is upheld, which is taken as evidence that the model provides a reasonably accurate description of the behavioural strategies that people are employing, the manner in which they make their decisions, and the particular cues to which they attend. Predictions can then be made about whether human beings in other environments might be expected to behave in a similar or different manner, and data from different populations can then be compared. Where the data do not fit, the model can be revised to include other variables and retested, if necessary, repeatedly, in a manner that allows researchers to refine their understanding of the particular behavioural strategy that they are investigating. If the model remains a poor match, the conclusion may have to be drawn that there is no evidence to support the idea that human beings are behaving optimally in that particular situation.

An example of a well-established theory that has been developed by behavioural ecologists is *optimal foraging theory* (Stephens and Krebs, 1986). Here, researchers construct models that specify a general set of decision rules for foragers as they search for food. A central assumption is that natural selection will have favoured foragers that make those choices that yield the greatest payoff, defined as the biggest difference between the benefits and costs of a particular strategy. Those foragers that consistently choose high-payoff strategies are expected to leave the greatest number of offspring, and these offspring are also likely to show a propensity for choosing this winning strategy. However, as the relationship between particular strategies and the number of viable offspring can be difficult to quantify, proximate currencies, such as the number of calories gained per hour of foraging, are used to estimate optimality, based on the assumption that those individuals that forage most efficiently by this measure will, on average, leave the most descendants.

Optimal foraging theory has been applied to the feeding behaviour of hummingbirds that extract nectar from flowers using their long tongues. One decision that a foraging hummingbird makes is how long to spend feeding at a particular flower before moving on to the next. Because the amount of effort required to extract the nectar increases as the flower becomes emptier, the birds are predicted to move to another flower in the clump, rather than extracting the final drops, earlier when the number of available flowers is high and to stay for longer when fewer flowers are available. As predicted, the empirical data showed that hummingbirds spent longer at each flower when flower density was low and when average travel costs between flower patches were correspondingly high (Pyke, 1978). While the foraging behaviour of hummingbirds is now known to be richer than predicted by these relatively simple models (Pritchard et al., 2017), much has been learned from the model-testing approach.

Optimal foraging theory has been applied to the feeding decisions of a broad range of animal species, providing novel perspectives on decision-making and behavioural strategies (King and Marshall, 2022). Which currency is being optimized by the animals is not always clear, so different models can be developed to predict behaviour if individuals are maximizing the rate of energy intake, compared with, say, minimizing time spent foraging or the risk of predation. In each case, how a currency is optimized may depend upon factors such as the type of food item chosen by the individual, the time spent foraging on a particular resource, the choice of foraging patch, and the decision to forage alone or with other individuals. Assumptions about the constraints that limit the animal's feasible choices, such as its sensory abilities and the distributions of resources in the environment, are also built into the analysis. The costs and benefits of choosing particular strategies are then computed and the optimal strategy derived. The predictive performance of different models can then be compared to assess their relative merits. In a later section, we will encounter an example of how optimal foraging models have been applied to the study of human foraging behaviour.

Adaptive trade-offs

In the real world, the effective harvesting of energy from food is not the only problem that animals have to solve, and the competing demands on their time and resources are frequently in conflict. The best feeding site may also be teeming with dangerous predators, or a location may be a good place for finding food but have nowhere suitable for a home. Also, a maximally effective diet may require animals to make a balanced choice of a range of prey items and not just grab as many calories as possible. Optimization models analyse how animals solve such trade-offs. Because any 'unit of effort' can only be invested once, evolution is thought to have shaped animals into strategists that optimize how this investment is allocated into different aspects of their life history (Stearns, 1992). Behavioural ecologists are interested in how environmental factors influence the costs and benefits of particular trade-offs. One example of a trade-off is between somatic effort (growth of body tissue) and reproductive effort; that is, to invest in oneself or to prioritize reproduction instead.

A second is between direct and indirect reproduction, that is, the trade-off between reproducing oneself and helping others to reproduce. A third is how to balance parental effort and mating effort, for example, through investing in current offspring or searching for new mates. A fourth trade-off is between investment in the quantity versus quality of offspring.

An example of an adaptive trade-off in birds is clutch size (i.e. the number of eggs laid per breeding attempt), where parents have to strike a balance between producing too many and too few chicks. In his classic books, *The Natural Regulation of Animal Numbers* (1954) and *Population Studies of Birds* (1966), zoologist David Lack described detailed studies that investigated whether individual birds produce optimal clutch sizes. Beyond a certain point, larger clutch size will result in lowered overall parental reproductive success; the number of chicks born increases, but their survival rate decreases, because the parents can no longer cope with feeding them all. In contrast, laying too few eggs represents a wasted breeding opportunity, if a pair could have raised more chicks with the resources available to them. Lack predicted that an intermediate clutch size would yield the highest lifetime reproductive success, in terms of successfully fledged chicks. His prediction was upheld in a population of great tits that he studied in Wytham Woods near Oxford, UK, as the average clutch size was lower than the maximum possible clutch size that females can lay and varied across years depending upon overall resource availability.

Simple models of optimality suggest that all individuals should aim for the same optimum, and hence variation in the behaviour of individuals may imply that some are behaving suboptimally. For example, birds that lay much larger than average clutches of eggs might have wasted their efforts by producing too many eggs. However, Lack (1954) pointed out that an individual's optimal strategy is likely to depend upon the resources available to that individual at that timepoint. Continuing with our example of clutch size, some parents may be more able to invest in their offspring because they are more experienced or live on better quality territories, as a result of which the optimal clutch size for these parents may be larger than the optimal clutch size for younger birds living on poorer quality territories. Therefore, although a trade-off between clutch size and number of surviving offspring is predicted for any specific individual, the expected negative relationship between clutch size and number of offspring may not be found when comparing across individuals, because the absolute amount of resources available to each will differ. An important limitation of using correlational data to determine trade-offs results from the presence of these correlations between behaviour patterns, physical traits, or life history parameters, often referred to as the problem of *phenotypic correlation*.

The problem of correlations between variables is highly relevant to studies of human beings and can be illustrated by a familiar example of human behaviour. When money is a limited resource, it can be spent on a car or a house, but not both, so for any given family there is a trade-off. However, a survey of the values of cars and houses across households will generally reveal a positive (rather than the expected negative) correlation between amount spent on cars and houses. This correlation is found because households differ in the total amount of money that they have to begin with, and those with more money generally choose to increase the amount spent on both of the two commodities. The equivalent outcome in our bird example is that parents on resource-rich territories will raise large numbers of surviving offspring. While human beings generally only give birth to one offspring at a time, in the next section we will see how the concept of trade-offs greatly influenced thinking about how human parents may optimally adjust their family size or the interbirth interval between children.

In the face of phenotypic correlation, researchers studying animal behaviour can test for evidence of trade-offs through experimentation. For instance, the trade-off between clutch size and offspring survival has been investigated experimentally in birds by altering the clutch size in nests in different territories. Such studies have shown that either increasing or decreasing clutch size reduces reproductive success (de Heij et al., 2006), supporting the idea that intermediate clutch sizes are optimal. However, as such experimental manipulation is not possible with human populations, care should be taken when interpreting correlations, or lack of correlations, between variables within a population. Researchers have attempted to partial out the effects of phenotypic correlations by controlling statistically for one or more confounding variables within a population (Sear, 2007). Alternatively, hypotheses can sometimes be tested by looking at comparisons between populations, rather than trying to explain variation within populations (Borgerhoff Mulder, 2000), while bearing in mind the potential pitfalls of extrapolating from between-individual to between-populations comparisons, and vice versa (Pollet et al., 2014). State-dependent models also allow sophisticated predictions to be made about how the optimal strategy may depend upon the state, or condition, of the individual.

Case studies

This section provides examples of research in Human Behavioural Ecology. We first describe the use of optimal foraging theory to study hunting behaviour in a non-industrialized human society, then explore the relationship between marriage practices and environmental conditions in a range of populations. We show how trade-offs between offspring number and offspring quality have been studied in humans, including in relation to the puzzle of the 'demographic transition'. Then, we discuss the idea that cooperative childrearing practices might underpin unique aspects of human life histories.

Foraging strategies and optical group size

The Inuit populations of Arctic Canada exploit a wide variety of animal species as food, ranging from fish and waterfowl to seals, caribou, and beluga whales. They also use a number of different hunting methods, including traditional forms such as hunting seals at breathing holes in the ice. Foraging is often a solitary activity, but hunting in groups is also a common occupation. Eric Alden Smith of the University of Washington, USA, was interested in whether he could gain an understanding of the foraging behaviour of the Inuit using an optimality approach. Smith (1985) posed questions such as 'Why do Inuit typically form groups of five to sixteen to hunt belugas and groups of two to ten to hunt seals, but hunt gamebirds alone?' An obvious answer would be prey size; however, whereas seals may occasionally be hunted by groups of one or two hunters, smaller lake trout are often hunted by ten individuals. One prediction from optimal foraging theory is that individuals will form groups of the size that maximizes the capture rate of prey. Smith therefore wondered whether foraging in groups may result in greater foraging success than would hunting alone, or in fact reduce an individual's hunting efficiency but provide other benefits, such as safety or exchange of information.

To test these predictions, Smith collected data on hunting group sizes and amount of prey caught, and investigated whether the most common group size for a particular prey type or hunting method was equal to the estimated optimal group size. His results were mixed. For prey that were most effectively caught by a single hunter, such as geese and ptarmigan, the most common foraging group size was indeed one. However, where more than one individual was generally required to capture a prey item, the most common group size was not equivalent to the predicted optimal. For example, hunts at seal breathing holes provided the best returns per hunter when the hunting group size was three, but the most common size was four and could range up to eight individuals. Smith concluded that either the net capture rate was a poor currency to use to estimate optimal group size or, more interestingly perhaps, individuals were unable to maximize their net intake when foraging involved social interactions. Imagine a hunter that wanted to try to catch a seal at a breathing hole. If the likelihood of catching a seal by hunting alone was very low, the hunter would attempt to join a hunting group to increase their chance of success. If a group of three hunters was about to leave on a foraging trip, the lone individual might try to join. Any other lone hunters would also probably want to combine with the group, as long as they would do better by joining than by hunting alone. If members of the group were not inclined, or were unable, to stop other hunters from joining, the size of the group would continue to grow until the individual payoffs of hunting in a group were about the same as for hunting alone.

Group size might grow still larger if the additional hunters were relatives or individuals that were learning to hunt. Hunting group size will therefore depend upon the relative costs and benefits both to the group members and potential joiners. This case study illustrates that when the data do not fit the assumptions of a simple model, the lack of fit can often be informative in pointing out what might actually be going on. Smith's finding—that the average group size for hunting seals at breathing holes was greater than his predicted optimum—supported the idea that optimization can be constrained by other factors such as social interactions and the strategies of others (Sibly, 1983). Strikingly similar results have been reported for hunting group size in lion populations, with group size being larger than predicted from an optimality model, as the benefits to an individual of joining a group are usually greater than the benefits of hunting alone (Mangel and Clark, 1988).

Theoretical models have provided Human Behavioural Ecologists with testable predictions about foraging behaviour, which remains an active area of research within this sub-field (Koster and Bird, 2024). Using predictions from optimal foraging theory, these researchers have asked, for example, what cues influence the decision to move between foraging patches or between camps (Pacheco-Cobos et al., 2019; Venkataraman et al., 2017), what rules explain the division and sharing of food resources (Jaeggi and Gurven, 2013), what effects the decision to hunt with dogs has on foraging strategies (Koster, 2008), and what might explain sexual divisions of labour and the predominance of hunting by men (Codding et al., 2011). In additional to optimal foraging theory, other models have been explored, such as the use of costly signalling theory to test whether hunting provides men with social attention and mating benefits (Stibbard-Hawkes, 2019). The strength of support for these models can only be tested by the continued collection of relevant data from human populations.

Marriage practices and subsistence patterns

Human Behavioural Ecologists have long been interested in whether human marriage patterns vary as a function of the local ecological context (Shenk, 2024). A rare form of human marriage that has received considerable attention from anthropologists is that of *polyandry*, in which one woman is legally married to two or more men. Polyandry has been particularly well studied in the Zanskar and Ladakh regions of Tibet, where a number of brothers may be married to one wife. In the Himalayan villages of these regions, montane desert surrounds small areas of land that can only be cultivated because of the streams of snowmelt that flow down from the glaciers. Communities therefore subsist on relatively small pockets of fertile land surrounded by arid mountain landscapes. Each family produces crops and raises animals on an estate that the eldest son usually inherits from his parents. The estate could be divided between the sons of the family; however, below a certain size, the divided farm would be too small to maintain a family. Instead, the brothers might benefit from jointly marrying a single wife and working the farm together.

John Crook and Stamati Crook of the University of Bristol and the University of Oxford, UK, respectively, documented the lives of these villagers and suggested that polyandry is functionally adaptive in the particularly harsh environment of these areas of Tibet (Crook and Crook, 1988). The researchers found evidence that women in polyandrous marriages had more offspring than women in monogamous marriages and thus apparently benefited from polyandry. However, did both brothers benefit from sharing a wife? Crook and Crook calculated that brothers would do well to remain in the polyandrous relationship, provided that they all had equal chance of fathering the offspring. This outcome was unlikely to occur, as eldest brothers usually benefited from seniority and had higher reproductive success than younger brother. However, Smith (1998) showed mathematically that younger brothers may nonetheless benefit from joining a polyandrous marriage with an older brother even where the paternity of offspring is skewed, for reasons similar to why it may benefit a lone individual to join a foraging group. Crook and Crook predicted that younger brothers should accept polyandry only when life circumstances constrained them to do so. Consistent with this prediction, when alternative sources of income became available, many younger brothers left to search for a wife of their own, and town-dwellers in the same regions of Tibet generally exhibited monogamy and divided any wealth between their children. The range of marriage practices in these regions indicates how individuals respond flexibly to the local circumstances.

Although polyandry remains a relatively rare marriage system (Starkweather and Hames, 2012), *polygyny*, where one man is married to more than one woman, is sanctioned in over 80% of societies that have been described in standard, cross-cultural databases. In polygynous societies, only those men with the most wealth or resources are generally able to support multiple wives, so the majority of relationships will actually be monogamous. The apparent benefit of polygyny for men is relatively well established, as polygynous men usually have more offspring than monogamous men. However, whether polygyny holds any benefits for women has remained less clear. Behavioural ecologists have suggested that a female may choose a suitor who already has a partner if he has more than double the resources of an unpaired bachelor. The idea that, over a certain threshold, the resources monopolized by a male will make him more attractive than a competitor with no other partners is referred to as the *polygyny threshold model* (Orians, 1969; Verner and Willson, 1966). Monique Borgerhoff Mulder of the University of California, Davis, USA, investigated whether, given the choice between marrying polygynously or monogamously, Kipsigis women in Kenya adopted the option that would lead to the greatest reproductive success. In this population, men can marry up to twelve wives, although two to four wives are a more common number. Wives and children are dependent upon the crops and animals produced on the husband's land, and the number of wives strongly correlates with a man's resource ownership. The data indicated that, given a choice of potential husbands that were either already married or unmarried, women married the man able to provide the most resources (Borgerhoff Mulder, 1990), upholding a prediction of the polygyny threshold model. However, Kipsigis women in polygynous marriages had fewer surviving offspring on average than monogamously married women. Similarly, in an Ethiopian population, although the first wives of polygynously married men had high reproductive success, second or third wives had fewer surviving offspring than monogamously married women (Gibson and Mace, 2007). Given the mixed evidence for a relationship between polygyny and child health outcomes (Lawson and Gibson, 2018), women may sometimes be constrained in their decisions of whether to join a polygynous marriage, particularly when marriages are arranged by partners and relatives. Yet, later wives, who are often older and from poorer families than first wives, may still be making optimal choices given their alternatives.

One of the challenges for the polygyny threshold model is to explain why monogamous relationships are the norm in agricultural societies (Ross et al., 2018), where resources are often monopolized by relatively small numbers of individuals. Historical data have shown that the transition from small-scale, horticultural practices to larger scale, agricultural production was associated with a more unequal distribution of wealth and stratified social systems. Why don't wealthy men in agricultural societies engage in polygyny? Early theorists suggested that powerful men in these societies might impose monogamy on the whole community in order to reduce conflict between men and encourage within-group cooperation, particularly during periods of warfare (Alexander, 1979). However, theoretical analyses by Human Behavioural Ecologists helped to explain this pattern (Ross et al., 2018). These models showed that, in agricultural societies, the majority of men will be relatively poor, and only a very small fraction of men will acquire sufficient wealth to support polygynous relationships, meaning that monogamy is the optimal strategy for the vast majority of the population. Even wealthy men might optimize their number of grandchildren by having a monogamous marriage and transmitting their wealth to a relatively small number of closely related children (Fortunato and Archetti, 2010). Thus, these studies of marriage practices highlight the complex effects that subsistence patterns can have on the decision-making of individuals and resultant patterns of marital partnerships.

Trade-offs between quantity and quality of offspring

Earlier we saw how clutch size in birds represents a trade-off between the number of eggs laid and fledgling survival rates. In human beings, which do not usually produce broods, the total number of offspring will be influenced by factors such as the length of time between births, known as the *interbirth interval*. Too short an interbirth interval can jeopardize the life of the younger infant, while a lengthy interbirth interval could represent a missed opportunity to reproduce. Nicholas Blurton Jones of the University of California, Los Angeles, USA, investigated whether the interbirth interval observed in a hunter–gatherer community, the !Kung San of southern Africa, optimized the number of surviving offspring. Blurton Jones (1986), who had studied bird behaviour with Niko Tinbergen at the University of Oxford, used the detailed demographic data collected by anthropologists Nancy Howell and Richard Lee to try to explain the relatively long interbirth intervals (around 4 years) in these communities.

In one of the earliest studies to apply optimality theory to human fertility, Blurton Jones and Richard Sibly (1978) modelled how the workload of the mother would influence the optimal interbirth interval. They predicted that short interbirth intervals would be accompanied by more offspring deaths and, using the data collected by Lee, indeed found that an interbirth interval much shorter than four years resulted in increased infant mortality (Blurton Jones, 1986). However, the opposite prediction would also have been plausible: offspring born after a short interbirth interval may have been *more* likely to survive than those born after a long interval because their mothers were in good physical condition, a case of phenotypic correlation. Blurton Jones acknowledged that the prediction of a positive correlation between interbirth interval and infant survival may have been overly simplistic but did fit with the data (Blurton Jones, 1997). Subsequent attempts to replicate these results were generally not supportive. For example, Kim Hill and Magdalena Hurtado at Arizona State University, USA, failed to show that shorter interbirth intervals were associated with lower reproductive success among the Ache of Paraguay (Hill and Hurtado, 1996). In this population, the body weight of the mother was a much more influential factor on infant survival than was interbirth interval.

In addition to the length of the interbirth interval and the physical condition of the mother, the wealth of the parents may influence family size. In a range of pre-industrial societies, wealthier parents are reported to have a larger average number of offspring than poorer families (Hopcroft, 2006). Ruth Mace, an anthropologist at University College London, UK, examined how wealth might influence reproductive decisions in a camel-herding Gabbra population in Kenya. In this population, wealth could be measurable in terms of the numbers of camels and goats owned by the family. When deciding whether to have another baby, Gabbra parents appeared to take into account the probability that they would be able to raise the child all the way through to marriage. In order to obtain a wife for their son, parents must give up part of their herd, as well as making a bridewealth payment. Having to support the marriages of too many sons could be of detriment to the rest of the family. Mace (1996) used a mathematical model to analyse the trade-off between family size and marriage prospects, and she showed that the decision of whether to have another child depended upon both the wealth of the family and the number of sons produced already. Families that already had sons and had limited wealth were most likely to limit their family size, suggesting that offspring quantity was being traded off against quality when resources were constrained. Trade-offs between offspring number and levels of parental investment in each offspring are not always straightforward (Lawson and Borgerhoff Mulder, 2016). For instance, in addition to the financial costs of childrearing, the costs and benefits of having children may be related to the amount of local family support (Sear et al., 2016).

Human Behavioural Ecologists have also explored the 'demographic transition', which refers to the historical shift from large families and high mortality rates to small families and low mortality rates that is seen in many industrialized societies. In societies that have gone through a demographic transition, wealth and family size are not always positively correlated. A positive relationship between these variables sometimes holds for men, but women frequently show the inverse pattern, with wealthier women having the fewest offspring (Stulp et al., 2016b). Why would people choose to limit their reproduction when resources are apparently plentiful? Behavioural ecologists have speculated that having fewer offspring, but investing high levels of resources in each one, could results in more descendants in the long term. However, analyses reveal little evidence that wealthy parents increase their number of grandchildren by having fewer offspring (e.g. Goodman et al., 2012). These data suggest that limits on family size might reflect the investment costs of rearing socially and economically competitive children (Lawson and Mace, 2011), plus the fact that children in post-transition societies contribute less to their own economic costs (Shenk et al., 2013). While the puzzle of the demographic transition remains to be fully resolved by behavioural ecology approaches, the concept of trade-offs between the quantity and quality of offspring provides a framework for asking such questions.

Cooperative childcare

Compared to other great ape species, human infants are weaned at a relatively young age, followed by a slow period of growth during childhood and a long overall period of dependency. Chimpanzees, in contrast, have infants that are weaned at a relatively later age but reach sexual maturity more quickly, and chimpanzee mothers generally raise one offspring to nutritional independence before having the next one. The fact that human mothers can produce infants in relatively close succession means that they can have multiple dependent offspring of different ages at one time. The energetic challenges of these overlapping childhoods can be substantial. In hunter–gatherer populations, for instance, children often do not produce enough calories through their own labours to cover their nutritional needs until adolescence at the earliest, particularly where strength and skills are required to extract those food resources (Urlacher, 2023). This unusual pattern of delayed independence has potentially been selected over evolutionary time periods because of the benefits of long periods of learning (Kaplan et al., 2000). But how do humans manage to raise such costly offspring? Human Behavioural Ecologists have proposed that the answer is *cooperative childrearing*, where a range of helpers provide care to offspring to alleviate the direct costs to the mother.

Cooperative care might come from a range of individuals within a mother's social network. One category of relatives who could particularly gain from helping out is maternal grandmothers. These grandmothers can be certain that the offspring of their daughters are direct descendants and, once these women have completed their own families, caring for grandchildren might have been selected through inclusive fitness benefits. The 'grandmother hypothesis' proposes that grandmothers play a significant role in raising grandchildren and that this additional care is key to the evolution of unique aspects of human life histories (Hawkes, 2003). In non-industrialized societies, maternal grandmothers have been shown to improve child survival rates and reduce spacing between births (Chapman et al., 2021; Sear and Mace, 2005), and, in industrialized contexts, maternal grandparents are particularly likely to direct practical and financial assistance towards their grandchildren (Pollet et al., 2009). However, siblings and other children also contribute substantially to childcare across societies (Kramer and Veile, 2018), starting at a relatively young age. For example, among the BaYaka of the Republic of Congo, young girls aged four to seven years help nursing mothers by holding, supervising, and playing with young infants, which allows mothers to engage in food gathering (Jang et al., 2022). Extra pairs of hands can help mothers to undertake activities that would otherwise be compromised by the constant demands of their children.

The most obvious caregiver to help a mother would be the child's father. In non-human primates, the level of investment by fathers is highly variable across species, ranging from no direct care (e.g. chimpanzees) to extensive carrying and food-sharing with their infants (e.g. owl monkeys; Rosenbaum and Silk, 2022). Compared to most other primate species, human fathers provide relatively high levels of investment in their offspring, yet the levels of paternal care vary greatly within and between human societies (Fouts, 2008). Even in societies with extensive paternal care, levels of caretaking by fathers are responsive to the availability of other caregivers (Sear, 2016). For example, among the Aka foragers of the Central African Republic, fathers who reside with their wife's family provide less direct care to their children than do those residing with the father's family (Meehan, 2005). This ability to switch facultatively between caregivers has led to the 'expendable male hypothesis', which posits that, where women live close to their relatives and where female control over resources is high, parental investment from fathers might not be required (Mattison et al., 2019). As well as relying on kin, women can also cooperate with non-related individuals, thus setting up a network of helpers. Among the Shodagor of Bangladesh, reciprocal cooperation between unrelated individuals is more important when the method of subsistence is incompatible with childcare (trading) than when tasks are childcare compatible (fishing) (Starkweather et al., 2023). These data highlight the flexibility and diversity of cooperative childrearing arrangements.

According to Human Behavioural Ecologists, the Western idea of a 'traditional' two-parent household, with a mother who looks after the children alone and a father who is employed outside the home, is now outdated (Sear, 2021b). Sarah Hrdy (2009) has argued that, throughout human evolutionary history, women have relied on an extensive network of alloparents to contribute to childcare, perhaps best summarized by the proverb 'it takes a village to raise a child'. Grandmothers have received particular attention as a key source of caregiving, but other members of social groups, including fathers, older siblings, and unrelated helpers, also provide significant levels of childcare. Theoretical research has suggested that provisioning by men is most likely when the diet contains energetically rich, but difficult-to-obtain, items, such as meat and animal products, leading to complementary foraging strategies between men and women (Alger et al., 2020). Yet the idea that 'man the hunter' was the most important character in human evolutionary history has been soundly critiqued (Lacy and Ocobock, 2024; Ocobock and Lacy, 2024). Humans have instead been described as 'cooperative breeders' (Hrdy, 1999, 2009), although 'communal breeders' is perhaps a more accurate term, given that, in non-human animals, cooperatively breeding species (e.g. marmosets, tamarins, and meerkats) exhibit physiological suppression of reproductive in subordinate individuals. In contrast, women who reside in close networks of female kin may be more likely to have children, rather than competing for reproductive opportunities (Hackman and Kramer, 2022).

Critical evaluation

In the late 1980s, Human Behavioural Ecology came under attack from the founders of the newly forming discipline of Evolutionary Psychology. The most hostile remarks came from Donald Symons (1987) at the University of California, Santa Barbara, USA, in an infamous essay entitled 'If we're all Darwinians, what's the fuss about?'. Symons argued that the research programme of the Human Behavioural Ecologists was seriously misguided because it did not formulate, or test, hypotheses concerning human adaptations or shed light on the human mind where such adaptations would be found. Symons proposed that natural selection does not act directly on behaviour, but rather on the behavioural regulatory machinery that underpins it. Confusingly, two very similar terms have been used for two distinct ideas. An

	Is the behaviour adaptive? *Adaptive behaviour* is functional behaviour that increases reproductive success.	
	Yes	No
Is the behaviour an adaptation? An *adaptation* is a character favoured by natural selection for its effectiveness in a particular role. **Yes**	**Current adaptation** A *current adaptation* is an adaptation that has remained adaptive due to a continuity in the selective environment.	**Past adaptation** A *past adaptation* is an adaptation that is no longer adaptive due to a change in the selective environment.
No	**Exaptation** An *exaptation* is a character that now enhances fitness but was not built by natural selection for its current role.	**Dysfunctional by-product** A *dysfunctional by-product* is a character that neither enhances fitness nor was built by natural selection.

Figure 4.1 The difference between adaptive behaviour and adaptations.

adaptation is a character favoured by natural selection for its effectiveness in a particular role; that is, it has an evolutionary history of selection. However, to be labelled as *adaptive*, a character only has to function currently to increase reproductive success. As Figure 4.1 illustrates, not only are adaptive traits not the same as adaptations, but they can be regarded as being independent. Human Behavioural Ecology was dismissed as merely establishing which behaviour patterns appear to be adaptive by correlating human behavioural traits with reproductive success in current environments, rather than identifying adaptations (Symons, 1987, 1989).

For Symons (1990, p. 430) 'correlating trait variation with reproductive success' was 'an ineffective, ambiguous, and inconclusive way to study adaptation'. He stressed how human characteristics that currently appear adaptive might not actually be adaptations. One reason for this discrepancy was pointed out by George Williams in his classic 1966 book *Adaptation and Natural Selection*. Williams emphasized how it is important to distinguish adaptations from characters with fortuitous effects. The latter, which Gould and Vrba (1982) subsequently termed 'exaptations', are features that now enhance fitness but were not built by natural selection for their current role. For instance, Williams wrote 'I cannot readily accept the idea that [human] advanced mental capabilities have ever been directly favoured by selection' although they 'might possibly be produced as an incidental effect of selection for the ability to understand and remember simple verbal instructions early in life' (1966, pp. 14–15). Here, language would be an adaptation, whereas human intelligence is an exaptation.

Symons maintained that many human adaptations might not be currently adaptive but were rather adaptations to a bygone world inhabited by our ancestors. For instance, whether or not it is currently adaptive, the human taste for sugar and fat in the diet is likely to be an adaptation to a past hunter–gatherer existence where our ancestors could not get enough of these energy-rich nutrients, and that is why these food items are still pleasurable to us in modern environments. Importantly, Symons also argued that the adaptations that underpin human behaviour were to be found at the psychological level: these adaptations were the cognitive machinery of the mind that controls behaviour and that was largely ignored by early Human Behavioural Ecologists. For Symons, the best way to use evolution to study human behaviour was to take an *adaptationist* approach; that is, to search for the psychological mechanisms that constituted the adaptations that regulate behaviour, rather than an *adaptivist* approach that identifies adaptive behaviour among humans but may bear no relationship to human adaptations:

> Darwinism is a historical explanation of the origin and maintenance of *adaptations*, and almost none of the phenomena of interest to social scientists—polyandry, bridewealth ... and so forth—are themselves adaptations. Whether or not they are adaptive, they cannot be adaptations because they are not descriptions of phenotypic design. Darwinism can be 'applied' to traditional social science phenomena only insofar as it illuminates the psychological adaptations that underpin those phenomena. (Symons, 1990, p. 435)

Other Evolutionary Psychologists, notably Leda Cosmides and John Tooby, joined in Symons's attack (Cosmides and Tooby, 1987; Tooby and Cosmides, 1990a). Unsurprisingly, the Human Behavioural Ecologists defended their position, and a vigorous, and sometimes bitter, debate ensued.

In the following sections, we examine the criticisms of Human Behavioural Ecology and evaluate whether these criticisms still apply to current research in this sub-field. More specifically, we examine the question of whether it is better to focus on human behaviour or psychological mechanisms and assess how the Human Behavioural Ecology approach deals with the possibility that human beings might sometimes exhibit suboptimal behaviour. We will then go on to consider a third line of criticism that emanates more from the social science community and questions the legitimacy of studying human behaviour and institutions in a piecemeal fashion.

Behaviour versus psychological mechanisms

Are Human Behavioural Ecologists justified in focusing on adaptive human behaviour? If not, are Evolutionary Psychologists, whom we will meet in the next chapter, correct in their claim that analysis of psychological mechanisms is the most appropriate level at which to look for human adaptations? Clearly there are strengths and weaknesses to both perspectives. The Human Behavioural Ecology position has the advantage that behaviour can be observed and quantified, so it can be studied in a rigorous scientific manner and can be modelled with formal mathematical theory. In reality, Evolutionary Psychologists also rely on observing behavioural traits and preferences, and usually infer the underlying cognitive processes rather than examining neural mechanisms directly. The Human Behavioural Ecology approach might not tell researchers much about human psychological adaptations, but this criticism has little traction when the goal of such studies is instead to understand the patterns of behaviour generated by the human mind. In terms of Tinbergen's four questions, Human Behavioural Ecologists have focused on asking one of these questions, namely what is the *function* of behaviour in relation to environmental parameters and social structure? In exploring what makes a behaviour adaptive, these researchers hope to pinpoint the function of the behaviour, and thereby shed light on why it evolved. The fact that other research fields might focus on alternative questions does not inherently detract from the goals of Human Behavioural Ecology.

Indeed, if different groups of researchers use alternative approaches, the topic can be examined from multiple angles, leading to a fuller understanding of the phenomenon. Human Behavioural Ecologists have been described as measuring the set of outcomes that are produced by the evolved decision-making processes that are resident in the brain (Kaplan and Gangestad, 2005). From this perspective, Human Behavioural Ecology and Evolutionary Psychology appear to be compatible, at least, in this respect, as information-processing algorithms in the brain could produce adaptive behavioural and life history responses to specific environmental cues. Alexander (1979) and Borgerhoff Mulder (1991) have argued that natural selection will choose between alternative sets of 'decision rules' and select those sets of rules that give the best outcome in terms of maximizing reproductive success. Human Behavioural Ecologists have thereby applied the concept of the *reaction norm*, a term that is used by biologists to describe the ranges of behaviour patterns, or variations in physical traits, exhibited by a particular genotype that result from exposure to different environmental conditions. Humans are thereby predicted to produce adaptive responses, as long as current environments are not too dissimilar from ancestral conditions.

Human Behavioural Ecologists do not generally place much emphasis on proximate mechanisms, such as whether variation in a particular behaviour is related to genes, psychological traits, or socially learned information, and many Human Behavioural Ecologists have generally regarded this agnostic position as a virtue. For instance, Smith (2000) was open about the Human Behavioural Ecologist's adherence to the *phenotypic gambit* (Grafen, 1984), which posits that the constraints on human adaptiveness can be ignored in the construction of models and the testing of hypotheses. The starting assumption is that human adaptiveness is only minimally constrained and that most proximate mechanisms do not significantly alter the optimal strategy or its environmental correlates (Mace, 2014). The researchers then compare the observed data with predictions made from this starting position and recursively revise the model to find the best fit between the model and data. As pointed out by Maynard Smith (1978), the role of optimization theory is not to demonstrate that the organism is behaving adaptively but to use the assumption of adaptive behaviour as a tool to develop an understanding of the diversity of behavioural strategies. How humans end up behaving in an adaptive manner is not necessarily relevant; as long as the behaviour is adaptive, it can be predicted with formal models. The key legacy of human evolutionary history is assumed to be *adaptability*, which must itself be an adaptation, albeit an extremely general one.

The idea that proximate mechanisms can be ignored when applying the concept of optimality has been challenged by some behavioural ecologists. For instance, Borgerhoff Mulder (2013) argued that simplistic behavioural ecology models, which fail to take into account the fact that human behaviour is strongly influenced by culture, might only take researchers so far in explaining behavioural variation. Understanding *how* human beings acquire behavioural traits might provide important insights into *why* individuals behave in particular ways. Animal behaviour researchers have made a similar case for greater integration of research on proximate mechanisms and the evolution of traits (McNamara and Houston, 2009). In recent years, Human Behavioural Ecologists have shown an increasing willingness to include questions about psychological traits and cultural transmission in their models (Sear et al., 2016), resulting in more complex models of human behaviour. For instance, small family sizes in industrialized societies are now thought to reflect a combination of prevailing environmental conditions, cultural and economic changes, and evolved aspects of psychology that influence decision-making (Colleran, 2016). In addition, a related research field that has always considered proximate mechanisms is *reproductive ecology* (Ellis, 2001). These researchers examine the relationships between human reproductive parameters, life history variables, and health, and they commonly incorporate questions about the role of reproductive hormones into their accounts of human behaviour and life history variation (e.g. Bribiescas, 2016; Jasienska, 2013). In

Chapter 8, we make the case that the amount of integrative research within the human evolutionary behavioural sciences has increased over time, with Human Behavioural Ecologists, in particular, forging links between their own discipline and adjacent sub-fields.

Human Behavioural Ecologists have also examined how culture itself might have been influenced by environmental parameters; for instance, the decline of the ancient Maya communities is thought to have occurred in response to extended periods of drought (Kennett et al., 2012). Human Behavioural Ecologists have also forged links with human demographers, who are interested in parameters such as births, deaths, and the timing of reproductive events. The resulting field of *evolutionary demography* has the potential to shed further light on the variation in human life histories within and between populations (Sear et al., 2024). The centrality of the phenotypic gambit to Human Behavioural Ecology research, and the neglect of mechanisms, including social learning, appears to be waning and is being replaced by a broader, multi-level approach to behavioural variation. While this integration of perspectives is surely welcome, it has not left the evolutionary behavioural sciences immune to criticism. Researchers have argued that the idea that behaviour largely consists of a set of predefined outputs that are 'programmed' during early life, or pre-specified by genetically determined reaction norms, does not place sufficient emphasis on the role of development in evolutionary processes. We return to this point in Chapter 8.

The possibility of suboptimal behaviour

Human Behavioural Ecologists have traditionally studied people living in small-scale, non-industrial societies, most likely for two main reasons. First, as most Human Behavioural Ecologists have a background in anthropology, these researchers will be familiar with the rich, fieldwork-based ethnographic literature on these societies and the diversity of their subsistence patterns, social organization, and cultural traditions. Second, Human Behavioural Ecologists reason that, although contemporary hunter–gatherers and foragers do not live in environments that are identical to those in which human beings evolved, these populations can provide insights into how our species may have adapted to ecological and social conditions in the past, where modern technologies and contraceptives will also have been absent (Borgerhoff Mulder and Schacht, 2012). Human Behavioural Ecologists are increasingly likely to test their hypotheses using data from non-traditional societies, as shown, for example, in the case study on early deprivation and life history milestones. Human Behavioural Ecology methods have therefore been successfully applied to a range of human populations.

Where behaviour appears to depart from optimality models, a number of explanations can be considered. One possibility is that the models have not taken into account all of the relevant variables or trade-offs with other aspects of behaviour or life history. The researchers can continue to update their models until the particular behaviour pattern can be explained. Cases in which the data do not fit the predictions of the model can be just as informative as cases in which a perfect fit is produced. For example, studies on the hunting behaviour of Ache men revealed that their behaviour did not fit that predicted by optimal foraging theory; they were spending more time hunting for meat items than expected. Kim Hill, based at Arizona State University, suggested that, as well as calorific content, measures of nutrient content needed to be included into the model for this population (Hill, 1988). Kristen Hawkes later considered the possibility that men were hunting more often than predicted to gain benefits in terms of matings (Hawkes, 1991). The data from Ache men did indeed suggest that the best hunters gained more extramarital matings than poor hunters (Kaplan and Hill, 1985). Thus, even when the models fail, light is frequently thrown on the phenomenon in question.

An alternative possibility is that human behavioural variation is not always adaptive. Evolutionary Psychologists have argued that, because modern environments are so different from those experienced by ancestral populations, researchers should not necessarily expect human beings to behave adaptively (Cosmides and Tooby, 1987). All organisms potentially exhibit some level of *mismatch* (or *adaptive lag*) between their evolved traits and current environments, particularly where the relevant environmental parameters have undergone recent change. Evolutionary Psychologists have argued that human culture and technology are particularly potent factors that have changed modern human environments rapidly and dramatically (Bowlby, 1969; Cosmides and Tooby, 1987). Mace agrees that 'culture might sometimes lead us astray from fitness maximisation' (Mace, 2014, p. 445). However, the extent to which human-constructed environments do generate mismatch is open to question. Humans are particularly adept at engineering their environments, largely through cultural processes, and natural selection is predicted to have favoured organisms that construct their environments in a manner that matches with their prior adaptations (Laland and Brown, 2006). This observation implies that the amount of mismatch may have been overestimated for human beings.

Nonetheless, human beings are not infinitely flexible and human-constructed environments are not universally benign. While the philosophy of Human Behavioural Ecology is to see how far the assumption that human beings behave optimally can be taken in terms of explaining the observed patterns of behaviour, many behavioural ecologists agree that proximate mechanism underlying specific behaviour patterns could potentially place constraints on the system and result in suboptimal behaviour (Fawcett et al., 2013; Nettle et al., 2013). For instance, the neural mechanisms that are involved in learning and decision-making have been selected over evolutionary time periods for their efficiency in particular ecological contexts, and these mechanisms can bias the expression of adaptive behaviour, resulting in apparently suboptimal choices being made in some instances. Brain evolution, like the evolution of all traits, involves a constant trade-off between efficiency and flexibility. Other behavioural ecologists have acknowledged genetic, developmental, and social constraints on animal behaviour (Kappeler et al., 2013). For instance, biologists have suggested that, although carnivores would perhaps benefit from the same behavioural and cognitive flexibility that is shown by primates, the functional specialization of the carnivore skull, jaw, and associated feeding muscles has impeded selection for advanced cognitive traits in this taxonomic group (Holekamp et al., 2013). Developmental mechanisms can also be expected to have imposed biases on human evolution.

Even when Human Behavioural Ecologists do find evidence that a behavioural trait fits with their optimality models, the trait might not be an adaptation per se. Evolutionary Psychologists (e.g. Symons, 1987; Tooby and Cosmides, 1990a) have pointed out the distinctions between adaptive behaviour, adaptations, and exaptations (see Figure 4.1). Evolutionary Psychologists are correct in stating that studying current function does not necessarily provide any information about whether that trait was brought about by selection for that particular function. However, if current selection acts on heritable variation in that trait, the trait will have an evolutionary present, which would technically transform an exaptation into an adaptation (Endler, 1986a). Human Behavioural Ecologists start with the premise that behaviour patterns may have a current adaptive function (two left-hand boxes, 'Current adaptation' and 'Exaptation', in Figure 4.1), although they need to remain aware of the distinction between current adaptations and exaptations, and of the existence of adaptations to past environments. In contrast, Evolutionary Psychologists are interested in the adaptations that influence human behaviour (and so concentrate on the top two boxes, 'Current adaptation' and 'Past adaptation', in Figure 4.1), but believe that many, perhaps even most, are no longer adaptive (past adaptations). Undoubtably, a complete evolutionary account would have to involve relating any observed adaptive or non-adaptive behaviour to the operation of an underlying adaptation and showing how the adaptation was adaptive in ancestral environments. Unfortunately, in the absence of detailed knowledge of our evolutionary past, that goal is not always easy to achieve. However, what happens in the present can inform our understanding of what happened in the past at least as much as vice versa, if for no other reason than that the present is more visible and more subject to experimentation (Turke, 1990).

Piecemeal approach

The anthropological community is dominated by holistic approaches to understanding human behaviour and has been critical of the piecemeal method adopted by the Human Behavioural Ecologists and other evolutionary minded researchers (e.g. Bloch, 2000). According to Smith (2000):

> The piecemeal approach holds that complex socioecological phenomena are fruitfully studied piece by piece ... Thus, a complex problem such as explaining the marriage patterns in a population is broken down into a set of component decisions and constraints such as the female preferences for mate characteristics, male preferences, the distribution of these characteristics in the population, the ecological and historical determinants of this distribution, and so on. (pp. 29–30)

How legitimate is it to assume that humans are optimizing only one aspect of behaviour and that complex behaviour can be analysed piece by piece? Are there not likely to be trade-offs between a number of important currencies, such as foraging success, social status, and mate choice, which might mean that studying only one aspect of behaviour would not be informative? Can human behaviour, social structure, and local traditions be simply broken down into separate units and measured?

One response is that the piecemeal approach is a necessary and pragmatic stance that can help researchers to understand complex phenomena. Interrelationships between human institutions, such as kinship, law, and religion, mean that numerous variables will need to be considered to gain a complete understanding of an individual's behaviour (Hinde, 1987), but it is rarely possible to investigate all these causal influences in a single study. In order to construct useful theoretical models, researchers need to make simplifying assumptions. Such deliberate simplification is usually a virtue, rather than a vice, to the extent that it focuses researchers' attention on the key processes that underlie the system and removes less relevant factors. Simple models can always be extended to relax their underlying assumptions, and analyses can be broadened to incorporate additional variables. Mathematical modelling, in general, is a dynamic process. The most effective approach is usually to start with a simple model that concentrates on the central processes and to elaborate gradually from that position. Human Behavioural Ecologists can point to the numerous examples where using such methods have provided novel insights into the behaviour, social organization, and life histories of people living in small-scale societies, including the Ache of Paraguay (Hill and Hurtado, 1996), Hadza of Tanzania (Marlowe, 2010), and Tsimane of Bolivia (Gurven et al., 2017).

Human Behavioural Ecologists have also shown that the piecemeal approach allows the use of sophisticated statistical methods that can shed light on how different aspects of behaviour relate to each other. For example, Clare Holden and Ruth Mace (2003) investigated the association between subsistence behaviour and inheritance patterns in a number of Bantu-speaking populations in sub-Saharan Africa, where family wealth is either passed down through the male lineage (patrilineal) or the female lineage (matrilineal). Their analyses showed that populations that engage in pastoralism and cattle herding tend to have patrilineal social systems, whereas matrilineal systems are more likely to occur in populations that don't engage in cattle herding. The researchers hypothesized that pastoralism allows men to monopolize resources and engage in polygynous marriages, which favours the transmission of wealth to sons. As these populations live in relatively close proximity to each other and might share traits simply because an ancestral population also exhibited these traits, the researchers used language 'trees' to estimate the shared ancestry of populations and then took this ancestry, or *phylogenetic relatedness*, into account in the statistical analyses. These types of cross-cultural analyses are only made possible by making simplifying assumptions, such as that behaviour patterns can be quantified as either being present or absent. Adopting these idealizations allows for insightful analyses that otherwise would be infeasible. Breaking behaviour down into components does not preclude exploring how those components interact; to the contrary, it facilitates that exploration.

One of the challenges for Human Behavioural Ecologists is to decide the direction of causality when two behaviour patterns occur together. Does behaviour pattern A account for the presence of behaviour pattern B, or does the causal relationship instead extend from B to A, or does a third variable C causally link both A and B? Researchers are able to use statistical approaches to ask precisely these types of question. For example, in the study of Bantu-speaking populations, the phylogenetic analyses supported the hypothesis that acquiring cattle led previously matrilineal communities to change to patrilineal or mixed wealth inheritance (Holden and Mace, 2003). Thus, by using data on the shared history of populations, researchers are able to ask whether it is more likely that one trait, such as an ecological variable or cultural trait, was present before the transition from one social system to another, or vice versa. Human Behavioural Ecologists are fully aware of the issue that correlations between variables do not necessarily imply causation (Mace and Jordan, 2011), but, equally, the claim that one cannot shed light on causality through correlational analyses is incorrect. Statistical methods now exist that allow researchers to translate a causal hypothesis into a corresponding model and thereby to distinguish between competing causal hypotheses using observational data (Shipley, 2016). In addition, ecological variables might not always be relevant in understanding the distribution of traits. For example, whether newly married couples live near the wife's or the husband's family is highly variable between closely related populations (Moravec et al., 2018) and is potentially related to culturally induced events rather than local ecology. Understanding causality is likely to require researchers to attend to the mechanisms underpinning the adoption of particular behavioural traits.

Taking a piecemeal approach does not mean that causality will always be easy to infer. For example, a study of historical records in Finland found that individuals that had experienced harsh environment conditions during early life, as measured by variables such as crop yields, had lower survival and reproductive success than individuals that experienced better conditions in early life when both groups were exposed to harsh conditions in adulthood (Hayward et al., 2013). The opposite pattern of results might have been expected, based on the idea of 'predictive adaptive responses', but the results might instead be an example of phenotypic correlation, whereby individuals in better condition have an advantage across all ecological parameters. Such correlations can obscure potential causal relationships between variables. In some instances, Human Behavioural Ecologists are able to take advantage of 'natural experiments' to test their causal relationships. For example, when water pump systems were introduced into villages in Southern Ethiopia, the researchers examined whether this intervention enhanced child and maternal health relative to villages without the water taps (Gibson and Mace, 2006). Consistent with the hypothesis that maternal health is traded off against investment in offspring, the data revealed that improving access to water reduced child mortality but did not have the predicted positive impact on maternal health, as birth rate also increased for the women in these villages. Based on this unexpected finding, Human Behavioural Ecologists could then refine their models to further understand fertility-related decision-making in this population. Taking a piecemeal approach, breaking down complex phenomena into quantifiable traits, allows causal relationships to be investigated.

Conclusion

Many anthropologists and other social scientists are sceptical about, if not downright hostile to, the evolutionary perspective of Human Behavioural Ecologists. Indeed, in many areas of the social sciences, 'post-modernist' thinking encourages an anti-science negativism. As we saw in Chapter 3, part of the hostility stems from an acute awareness of the past abuses of Darwinism. Once bitten, the field of social anthropology has remained shy of evolutionary reasoning. Thus, while the methods of Human Behavioural Ecology have the advantage that they are quantitative, rigorous, theory driven, and insightful, such qualities are not always appreciated by those in the anthropological community at large, few of whom have a mathematical training. As a consequence, despite the rich vein of empirical evidence and insights that have emerged from Human Behavioural Ecology (Koster et al., 2024), and which are manifest in several hundred scholarly publications, the approach remains a relatively small branch of anthropology. Part of the challenge for the sub-field perhaps lies in the complexities of collecting long-term, individual-based datasets that generally underpin the hypothesis testing (McElreath and Koster, 2024). Mace (2014) has argued that consilience between social anthropologists and evolutionary-minded anthropologists might no longer be possible, given the current distance between these sets of researchers. Instead, at least some Human Behavioural Ecologists have nurtured positive connections with Cultural Evolution theorists (see Chapter 6) and are increasingly considering the role of culture in their models (Borgerhoff Mulder et al., 2009; Smith, 2013). In Chapter 8, we will examine the extent to which these two perspectives are fully compatible.

Human Behavioural Ecology has also been linked with the sub-field of Evolutionary Psychology, which we explore next, in Chapter 5. In terms of the number of researchers, Human Behavioural Ecology is dwarfed by its cousin Evolutionary Psychology. As we have seen, these two evolutionary approaches to studying human behaviour have sometimes been involved in heated debates, for instance, about the relative importance of studying behaviour versus psychological mechanisms, although the suggestion that Human Behavioural Ecologists may be documenting the behavioural outputs of evolved psychological mechanisms appears to have resolved this dispute to some extent (Barrett, 2024; Kaplan and Gangestad, 2005). The sub-field of Human Behavioural Ecology arguably provides a core component of any evolutionary approach to the study of human behaviour by providing a set of theoretical and methodological tools for examining reproductive success or its proxies in a diverse range of populations. Yet some key questions remain in terms of its coexistence with other sub-fields. Does the human mind contain predefined sets of behavioural outcomes that are 'evoked' by local environmental conditions, or are 'programmed' by early life conditions? Does the 'reaction norm' conception provide an adequate explanation of human behavioural development and diversity? What role does culture play in the evolution of human behaviour and our psychological traits? To answer these questions, we first need to take a closer look at the sub-field of Evolutionary Psychology.

5
Evolutionary psychology

Researchers committed to an evolutionary perspective on human behaviour were initially united in the face of widespread hostility to human sociobiology. However, in the 1980s, as the number of investigators using evolution to study human behaviour increased, subgroups began to emerge with different opinions on how best to proceed. One such subgroup was dominated by academic psychologists searching for the *evolved psychological mechanisms* that they envisaged underpinned the universal behavioural characteristics of humanity. Although the intellectual roots of some of these practitioners could be traced to human sociobiology or to the study of animal behaviour, the majority were fresh recruits who sought to differentiate themselves from human sociobiology and restyled themselves as Evolutionary Psychologists (henceforth, we capitalize this phrase to denote the specific sub-field that is characterized by the key concepts described below, as opposed to using this term to refer any psychologist who takes an evolutionary approach).

For Leda Cosmides and John Tooby, two of the pioneers of this new subfield, Evolutionary Psychology owed little intellectual debt to Edward Wilson's version of sociobiology but did draw inspiration from the writings of William Hamilton, Robert Trivers, and George Williams. Tooby, a Harvard-trained anthropologist who had worked closely with Irven DeVore, and Cosmides, a psychologist also from Harvard, were brought by Donald Symons to the University of California, Santa Barbara, USA, where they jointly founded the first Center for Evolutionary Psychology. The 'Santa Barbara school' was concerned that human sociobiologists and Human Behavioural Ecologists had neglected psychological adaptations:

> In the rush to apply evolutionary insights to a science of human behavior, many researchers have made a conceptual 'wrong turn', leaving a gap in the evolutionary approach that has limited its effectiveness. This wrong turn has consisted of attempting to apply evolutionary theory directly to the level of manifest behavior, rather than using it as a heuristic guide for the discovery of innate psychological mechanisms. (Cosmides and Tooby, 1987, pp. 278–279)

For Evolutionary Psychologists, any failure on the part of these other researchers to find evidence that human behaviour is 'optimal' could be explained by the fact that they were working at the wrong level of analysis by not considering the psychological mechanisms that underpin behaviour (Symons, 1987).

In addition, early Evolutionary Psychologists stressed that the physical and social environments experienced by contemporary human populations differ substantially from those experienced by our ancestors. Modern houses, cities, and social institutions are relatively recent innovations in evolutionary terms. Hence, Evolutionary Psychologists suggested that there is typically a mismatch, or 'adaptive lag', between our ancient psychological adaptations and our modern, constructed world. Because of this mismatch, they argued that researchers should not necessarily expect current human behaviour to be adaptive. Instead, if the human mind is comprised of many past adaptations, then the manner in which people think should reflect their ancestral selective environments. Evolutionary Psychologists proposed that evolutionary biology is best used to generate hypotheses about the adaptive problems that the human mind had to solve in the selective environment of our ancestors. Following John Bowlby (1969), this past environment was termed the *environment of evolutionary adaptedness* (EEA) and was described as the environment inhabited by our Stone Age hunter–gatherer ancestors during the Pleistocene.[1] Evolutionary Psychologists argued that researchers could make sense of the design features of the human mind by considering how our psychological processes evolved to solve problems encountered in the EEA. This theory-driven approach would be used to develop models of how the mind works.

The Evolutionary Psychologists' approach was also influenced by broader developments within the field of psychology. By the 1980s, *behaviourism* had become less fashionable, and psychology was in the throes of the *cognitive revolution*. This new perspective focused on describing the brain in computational terms. The use of animals as research tools was demoted in favour of using the computer as an analogue of human cognition. Minds were now described in terms of information processing and decision rules. For example, information received from sensory inputs was assumed to be used to generate representations of the external world, with cognitive decision-making processes then determining the relevant motor outputs. Psychologists were developing computational theories of informational processing that specified what had to happen if a particular function was to be accomplished (Marr, 1982).

---

[1] The Pleistocene is the geological time period from 2,580,000 years Before Present (BP) to 11,700 years BP.

Early research into artificial intelligence revealed that, to solve even supposedly simple cognitive tasks, minds required pre-specified procedures or information, and breaking problems down into simple functional tasks appeared critical to producing accurate simulations of intelligent behaviour.

This framework led Evolutionary Psychologists to propose that 'innate psychological mechanisms' guided decision-making and were organized into functional subroutines (Tooby and Cosmides, 1992). Evolutionary Psychologists believed that, with sufficient information about our ancestors' way of life, evolutionary theory could be used to construct computational theories of adaptive information processing. Cosmides and Tooby's visionary writings were to provide the defining features of the field, triggering rapid growth of this new movement. By the 1990s, Evolutionary Psychology had blossomed into a thriving research programme, as marked by the publication of Jerome Barkow, Cosmides, and Tooby's (1992) landmark volume, *The Adapted Mind*. The field was further bolstered by the success of a stream of popular science books, including Steven Pinker's (1997) *How the Mind Works*.

Tooby and Cosmides (1989a) outlined various steps that researchers must undertake to carry out Evolutionary Psychology research. These steps included using evolutionary theory as a starting point to develop models of adaptive problems that the human mind had to solve, determining how these adaptive problems manifested themselves in ancestral conditions, and then characterizing the design features that any cognitive programme must have had to solve these problems. Finally, the researcher could confirm expectations and eliminate alternative models using experiments and behavioural observations. As an illustration of how these methods could be put into practice, Tooby and Cosmides (1989b) considered social exchange: they began with Hamilton's (1964) inclusive fitness theory, which predicts that individuals ought to be most likely to behave altruistically towards close kin, and went on to postulate that cooperative exchanges between closely related members of a foraging band might have been critical for survival among our Pleistocene ancestors. For humans to confer such benefits on kin, they must have required cognitive programmes that allowed them to determine what cues were reliably shared by relatives, which led to the inference that humans will possess psychological mechanisms that allow them to extract this information and thus recognize their kin. Finally, experimental techniques were used to examine the features of human kin detection systems (Lieberman et al., 2007).

An additional strategy for generating and testing Evolutionary Psychology hypotheses was outlined by David Buss (1999) at the University of Texas, USA. Buss' observation-driven approach required researchers to develop a hypothesis about the adaptive function of a trait based on observations of behaviour or cognitive performance, and then to test further predictions about the functioning of the human mind based on this hypothesis. In other words, researchers start with the end-product and attempt to reconstruct the steps that led to this outcome, which is why this approach was referred to as 'reverse-engineering' (Pinker, 1997). Evolutionary Psychologists have subsequently embraced a broader range of strategies for generating their hypotheses, including the use of cross-species comparative approaches (Buss, 2015; Shackelford, 2020). Their methods have also broadened out from traditional psychological tools, such as laboratory experiments and questionnaires, to include field observations and the analysis of public records (Buss, 2019). In this chapter, we examine to what extent the Evolutionary Psychology approach has expanded the understanding of how evolutionary processes have shaped the human mind.

Key concepts

Three distinctive theoretical concepts of Evolutionary Psychology are that evolved psychological mechanisms underlie human behaviour, that the concept of the *EEA* can help researchers to delineate these psychological mechanisms, and that evolution has generally favoured *domain-specific* mental adaptations, or *modules*, that were adept at solving specific computational problems faced by our ancestors.

Evolved psychological mechanisms

An overarching goal of Evolutionary Psychology is to understand how the human mind is organized and to identify which human psychological traits are adaptations. Cosmides and Tooby (1987, p. 281) contend that 'natural selection cannot select for behavior per se; it can only select for mechanisms that produce behavior'. An *evolved psychological mechanism* is the term they gave to such mental adaptations, the information processing circuits in our brains that shape behaviour and cognition. Subsequently, the definition of psychological mechanisms was broadened to include emotions, preferences, and proclivities, and to include the concept of 'reaction norms', whereby psychological mechanisms can produce a range of behavioural outputs depending upon the specific context that the individual is experiencing (Barrett, 2015; Buss, 2019; Cosmides and Tooby, 2000). Psychological mechanisms are assumed to have evolved at a relatively slow pace and to exist in the form that they do because they recurrently solved a specific problem of survival or reproduction over a prolonged evolutionary history. According to Buss, who was paraphrasing the Greek philosopher Plato, evolved psychological mechanisms provide non-arbitrary criteria for 'carving the mind at its natural joints' (2008, p. 53). Buss envisaged that the mind possesses hundreds, perhaps even thousands, of evolved psychological mechanisms that are assumed to be universal (i.e. shared by all human beings), or at least relatively stable, across all human beings.

Sexual jealousy is said to be an example of an evolved psychological mechanism (Buss, 1994). In ancestral environments, males perhaps experienced jealous emotions when they observed their female partner behaving in an overly friendly manner to another male, and those males that were spurred into action as a consequence may have had a selective advantage over males who didn't experience this same emotional response. How each male went about expressing his jealousy might depend on factors such as his personality, his size, the size of the rival, and so on. One male might respond with threats or aggression towards the other male, another with increased vigilance, and another with an increased effort to fulfil his partner's desires. At the behavioural level, it is difficult to predict how an individual might respond to such situations, or why some instances even escalate to homicidal behaviour. However, Evolutionary Psychologists predict that past selection has favoured the evolution of jealous emotions, so, at the psychological level, a reliable pattern is found (Buss, 2018). Other emotions that have been described as evolved psychological mechanisms include gratitude towards others and a fear of snakes and spiders (McCullough et al., 2008; Öhman and Mineka, 2001).

For Evolutionary Psychologists (e.g. Tooby and Cosmides, 2005), the promise of the evolutionary perspective lies in its power to assist in the discovery, inventory, and analysis of the psychological mechanisms that underpin 'human nature'. Evolutionary Psychologists have thus placed emphasis on a species-specific repertoire of universal evolved psychological mechanisms, while also exploring how these mechanisms lead to behavioural variation. Anthropologist Donald Brown (1991) documented some of the traits that appear to be human universals. For instance, he reports how all people have a spoken language, and that all of these languages have phonemes, morphemes, and syntax; all societies are structured by status, roles, and a division of labour; and all possess incest avoidance regulations. Universal psychological mechanisms can potentially lead to diversity in human behaviour through *context-dependent strategies*, whereby local environmental variables act like 'switches' and shift the behavioural output of this universal, genetically based programme in an adaptive manner (Gangestad and Simpson, 2000;

Tooby and Cosmides, 1992). A metaphor that is commonly used to explain this behavioural switching is a musical jukebox that contains a large number of vinyl records: the evolved structure of the mind responds to different environmental conditions with alternative behaviour in a similar way to how the machinery of a jukebox responds when its buttons are pressed to change the tune that it plays.

Humans possess universals of behavioural development (Bateson and Martin, 1999). For example, with some exceptions, all human beings pass through roughly the same sequence of developmental milestones as they grow up, with most children starting to walk at 18 months and to talk at 2 years, and with most reaching sexual maturity by their late teens. *Evolutionary Developmental Psychologists* have proposed that the expression of evolved psychological traits changes over the course of development (Bjorklund et al., 2015). For example, certain psychological mechanisms, such as those that underpin childhood play, may only be relevant at a particular stage of development. In addition, early developmental environments, such as exposure to stress, are suggested to have long-term effects on behaviour by altering risk-taking, defensive reactions, or social affiliation (Del Giudice et al., 2011). Therefore, although the major focus of Evolutionary Psychology has been on universals, this perspective has widened out to account for individual-level variation in how putative evolved psychological mechanisms are expressed (Bjorklund, 2015; Mitchell, 2018). From an Evolutionary Psychology perspective, this variation in psychological traits is neither random nor infinite, given that both genetic and developmental constraints are assumed to direct the expression of evolved psychological traits towards outcomes that had adaptive benefits in ancestral environments, even though these outcomes may be maladaptive in the modern world.

The environment of evolutionary adaptedness

The concept of the *EEA* was initially developed by the British psychiatrist John Bowlby (1969), who was mentioned in Chapter 2, to explain why young children develop a strong attachment to their mothers, and why separation can result in extreme distress, including psychiatric disorders. Bowlby argued that the strong attachment of young to their parents should not be regarded as an illness or as dysfunctional behaviour, but rather as an adaptation that, in our evolutionary past, greatly enhanced the survival prospects of infants. Bowlby asserted that people have lived in modern societies with agriculture, high population density, and complex social institutions for only a few thousand years, whereas their predecessors lived in small foraging societies for a much longer time period. The modern world is therefore very different from that experienced by our genus, *Homo*, for most of its two-million-year history. Bowlby envisaged that separation anxiety in children was valuable when, and where, the behaviour evolved. The 'environment of evolutionary adaptedness' (henceforth EEA) is the term that Bowlby gave to this past selective environment. Prior to the 1960s, there was much confusion over the use of the term 'adaptation', and Bowlby's point that evolved characters may be adaptations to past environments was of considerable value.

In their writings on Evolutionary Psychology, Cosmides and Tooby rapidly adopted Bowlby's notion of the EEA. They also stressed how modern culture can change extremely quickly compared to the rate of biological evolution, potentially leaving our evolved psychological mechanisms lagging behind (Cosmides and Tooby, 1987). They reasoned that, if researchers could establish what kind of problems our Stone Age ancestors faced, they might be able to predict the kind of psychological mechanisms necessary to solve these problems, and hence which traits may be expected to have evolved. Early Evolutionary Psychologists asserted that the human mind was fashioned over the last two million years for a past world of hunting and gathering on the African plains of the Pleistocene epoch. For instance, Cosmides and Tooby wrote:

> Our species spent over 99% of its evolutionary history as hunter–gatherers in Pleistocene environments. (1987, pp. 280–281)

This statement does not negate the fact that every human descends from a lineage of ancestors that spans back through the billions of years of life on Earth. Although much of the social behaviour that human beings share with great apes, such as developing dominance hierarchies, understanding third-party social relationships, and coordinated hunting, probably evolved in our prehominid primate ancestors, many of our perceptual and learning abilities are likely to be phylogenetically older (van Duijn, 2017).

Tooby and Cosmides later clarified their position by stating that:

> [The EEA concept does not refer to a single] place or habitat, or even a time period. Rather, it is a statistical composite of the adaptation relevant properties of the ancestral environments encountered by members of ancestral populations, weighted by their frequency and their fitness consequences. (1990a, pp. 386–387)

They have also suggested that findings from palaeoanthropology, archaeology, and ethnographic studies of contemporary hunter–gatherers can be triangulated to establish robust features of our ancestors' life histories (these distant ancestors were omnivores, were exposed to diseases and predators, exhibited biparental care of offspring and sexual division of labour, engaged in cooperative exchanges, etc.) (Tooby and Cosmides, 2005). For Evolutionary Psychologists, this level of knowledge can potentially be sufficient to generate hypotheses about evolved psychological mechanisms. Although Evolutionary Psychologists do not always explicitly refer to the EEA in their writings, the suggestion that human beings experience an 'adaptive lag', or 'mismatch', between our psychological mechanisms and current environments remains a core assumption within this sub-field (e.g. Goetz et al., 2019; Li et al., 2018). The idea that our evolved psychological mechanisms are particularly suited to our past selective environments, and that these complex mechanisms evolve too slowly to have undergone significant change since the end of the Pleistocene, permeates this research tradition.

Domain specificity

A key foundation of Evolutionary Psychology is that minds are composed of a large number of psychological mechanisms dedicated to finding quick and efficient solutions to particular problems that were of significance to our ancestors. These psychological mechanisms are believed to have evolved to operate in a specific domain; in other words, the mechanisms are specialized, or *domain specific*. Commonly referenced *domains* include language, mate choice, sexual behaviour, parenting, friendship, resource accrual, disease avoidance, and social exchange. This standpoint can be contrasted with that of many cognitive scientists and neuroscientists, who view the human mind as being characterized by its plasticity and adaptability, and therefore heavily reliant on processes that operate across multiple domains; that is, the brain utilizes *domain-general* mechanisms. While artificial intelligence researchers are striving to recreate the domain-general processing abilities of real-life organisms (Roli et al., 2022), Evolutionary Psychologists regard an over-reliance on domain-generality as infeasible. According to Buss, evolved psychological mechanisms tend to be problem-specific because:

(1) general solutions fail to guide the organism to the correct adaptive solutions;
(2) even if they do work, general solutions lead to too many errors and thus are

costly to the organism; and (3) what constitutes a 'successful solution' differs from problem to problem. (2008, p. 54)

Likewise, Clark Barrett, at the University of California, Los Angeles, USA, maintains that 'all adaptations are domain-specific' (2015, p. 27), although he adopts a more flexible notion of domains. Evolutionary Psychologists suggest that humans have evolved specialized cognitive mechanisms that sort experience into adaptively meaningful channels that focus attention, organize perception and memory, and call up specialized procedural knowledge that will generate appropriate inferences, judgements, and choices given the context (Cosmides and Tooby, 1987). In this respect, the mind has been described as a 'Swiss army knife', with each psychological mechanism being analogous to a particular blade or tool.

In making the argument that psychological mechanisms are domain-specific, Evolutionary Psychologists draw on evidence that non-human animals are predisposed to learn some things and not others. For example, John Garcia, a psychologist at the University of California, Berkeley, USA, conducted early experiments in which rats were given a type of food and then, after several hours, a dose of radiation that made them sick. He found that the rats tended subsequently to avoid this specific type of food, and they did so because they had learned, often after just a single trial, that food with that particular taste led to illness. However, the rats struggled to learn an association between the other characteristics of the food and feeling sick, and they were extremely slow to learn that a buzzer sound or a light predicts illness (Garcia and Koelling, 1966). From an evolutionary perspective, these results make sense, as sickness generally results from eating rather than from noises or lights, and taste is a reliable indicator of a food's characteristics. Other psychologists showed that monkeys could acquire a fear of snakes after witnessing the fear responses of conspecifics but failed to learn a fear of an arbitrary stimulus (flowers) that way (Cook and Mineka, 1989). These experiments suggested that animals, including human beings, are *prepared* by evolution to learn some things more easily and quickly than others.

Case studies

In this section, we present four case studies that illustrate the Evolutionary Psychology approach. We first describe experimental evidence of a psychological mechanism for detecting cheating behaviour during social interactions, and then examine studies of human mating preferences. We go on to look at an evolutionary analysis of homicide and finally at studies of disgust responses.

Psychological mechanisms for detecting cheaters

Evolutionary Psychologists have reasoned that, if reciprocal altruism has been important in our evolutionary past, human beings should possess psychological mechanisms that render them sensitive to cheating during social interactions. In other words, human beings should be particularly good at detecting when individuals take the benefits from a social exchange without paying the costs. One widely used experimental paradigm for exploring how human beings carry out such analyses has been the 'Wason selection task'. British psychologist Peter Wason (1966) set out to understand human reasoning by devising an experiment to determine whether participants are able to detect violations of 'conditional rules'. An example of a conditional rule is that 'If you take the benefit, then you must pay the cost', and these rules can usually be represented in abstract terms, such as 'If P, then Q'. Wason found that people do not use logical reasoning all of the time and that the subject matter people are asked to think about affects how well they do on these tests. As an example, consider the task to detect violations of the abstract rule 'If a person has a "D" rating, then the document must be marked code "3" depicted in Figure 5.1a. Wason found that typically less than 25% of participants answer this task correctly. You can try this test for yourself before reading what the correct answer should be.

Most participants presented with this abstract problem said that only the D card, or the D and 3 cards, needed to be turned over to check for violations of the rule. In fact, the right answer is to turn over the D and 7 cards because, to establish that every D card has a 3 on the flip side, it is clearly necessary to turn over the D card, but it is also important to establish that the 7 is not a D. Whether the 3 is a D or not is irrelevant, as the rule does not insist that D is the only rating with the code 3. Now compare your performance with the task shown in Figure 5.1b. Surprisingly, despite the fact that the drinking-age task depicted has the same conditional rule as the abstract task, people consistently perform better on this task, with approximately 75% of subjects giving the logically correct response of 'drinking beer' and '16 years of age'. In both tasks, individuals are given a conditional rule of the form 'If P then Q' (i.e. if D then 3, or, if 'beer' then 'at least 21 years of age'), then asked what they need to do to determine whether this rule has been violated. The rule is violated only when

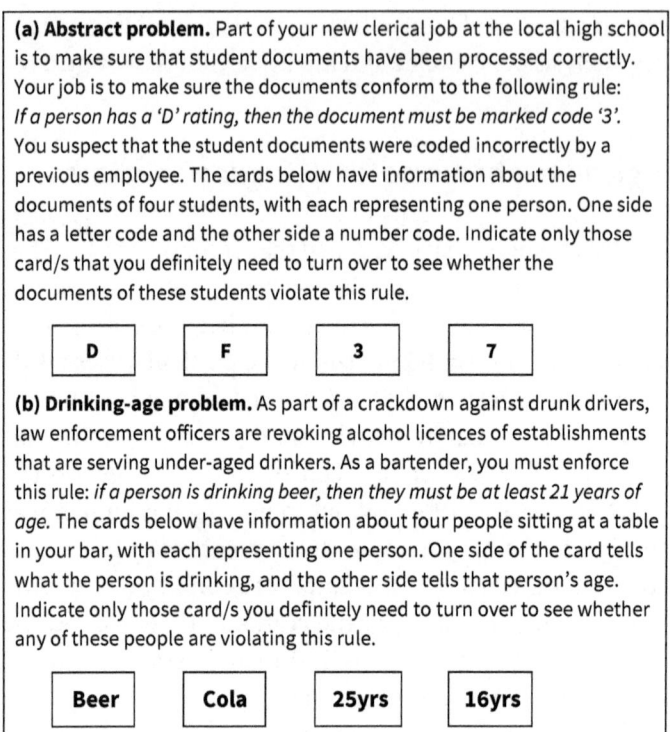

Figure 5.1 (a) Abstract problem and (b) drinking-age problem. Adapted from Tooby and Cosmides (1992) in J. H. Barkow, L. Cosmides and J. Tooby (eds), *The Adapted Mind: Evolutionary Psychology and the Generation of Culture*, with permission from Oxford University Press.

P is true and Q is false, and thus in both cases the answer is to check 'P' (the D or beer card) and 'not Q' (the 7 card or 16-years card).

Such experiments suggest that human reasoning changes depending on the subject matter about which one is reasoning, but no satisfactory theory could account for these content effects. As part of her doctoral research at Harvard University, Leda Cosmides set out to establish whether the findings made sense in evolutionary terms. In particular, Cosmides proposed that a history of reciprocal altruism among our ancestors has fashioned us with a 'cheater detection mechanism' that biases our reasoning. In a series of experiments that expanded Wason's findings, Cosmides found that, when subjects are asked to look for violations of conditional rules that relate to social, rather than non-social, contracts, their performance improves dramatically (Cosmides, 1989). According to Cosmides, the reason most people get the drinking-age task correct, but the abstract problem incorrect, is that only in the former case does the task implicitly involve cheater detection (Cosmides and Tooby, 1992). The drinking-age task has a content equivalent to 'If you take the benefit (drinking beer), then you pay the cost (being over 21)'. Underage drinking therefore violates a social norm and activates our ability to detect cheaters in social scenarios.

Cosmides' experiments ruled out alternative explanations, such as that performance was better on tasks where the content was more familiar. Even when participants were given an unfamiliar social norm, such as 'If a man eats cassava root, then he must have a tattoo on his face', they responded with a high level of success, provided that the preamble gave them sufficient information to establish that the rule was a social contract. In addition, Cosmides was able to switch the rules so that the logically correct answer conflicted with the social contract, and participants still responded in a manner consistent with the cheater detection hypothesis (Cosmides and Tooby, 1992). However, not all researchers accepted this interpretation of the Wason selection task; for example, some argued that the statements used in the different tasks are not identical logical forms and that the results depend upon how participants interpret the task (Buller, 2005; Sperber et al., 1995). Cosmides and colleagues responded with further studies (Cosmides et al., 2010; Fiddick et al., 2000), which they argue support their interpretation. Irrespective of which position is correct, few would dispute that Cosmides' experiments reinvigorated this topic. For instance, researchers have investigated whether participants are also good at detecting violations of rules that relate to maintaining coalitions and respecting social hierarchies (Sivan et al., 2018), whether human memory is biased towards remembering the faces of those who have cheated (Bell and Buchner, 2012), and whether a cheater detection mechanism might underpin human susceptibility to believing conspiracy theories (Aubert-Teillaud and Vaidis, 2024).

Sex differences in mate preferences

As natural selection operates through the differential reproduction of individuals, any psychological mechanisms that guide reproduction should be especially strong targets of selection. As a consequence, courtship and sex have been a major focus of Evolutionary Psychology. One question that has received considerable attention is whether evolution has fashioned us with preferences for particular characteristics in the opposite sex that influence our choice of mating partner. Based on Darwin's theory of sexual selection, Trivers (1972) proposed that, in animal species where females undertake the majority of parental care, females should exhibit a preference for males who are able to invest resources that would contribute to raising these offspring, such as food, shelter, and protection. In contrast, males are predicted to place greater emphasis on the 'reproductive value' of potential partners, which is related to physical fitness, age, and social rank, as a means of maximizing reproductive success. In addition, based on studies of fruit flies (Bateman, 1948), evolutionary biologists have reasoned that, in many species, males will be selected to mate with several partners as a means of enhancing their reproductive success, while females might be expected to favour fewer partners, as there is a limit on the maximum number of offspring that females can produce. The application of these ideas to studies of non-human animals has provided novel insights into the evolution of reproductive strategies across numerous species (Zuk and Simmons, 2018).

Perhaps a similar argument could also be applied to our own species. Evolutionary Psychologists have reasoned that, from the perspective of our ancestors in the EEA, women faced the burdens of internal fertilization and a nine-month gestation, followed by a lengthy period of lactation, and would consequently have benefited from selecting mates who possess resources (Buss, 1994). These researchers have also suggested that women might be selected to favour men who display cues indicating their resources, such as wealth or status, or their potential to accrue substantive resources in the future, for instance, through intelligence, hard work, and ambition. In contrast, men have been predicted to place greater emphasis on the physical attractiveness and youthfulness of potential partners (Buss, 1994). To test these hypotheses, Buss and his collaborators surveyed over 10,000 people in over thirty countries, asking them to rank the relative importance of a set of eighteen specific characteristics in potential romantic partners (Buss et al., 1990). As they predicted, the results revealed that, when the data were averaged across all nations, women placed more emphasis than men on the financial prospects of potential partners, whereas men placed more value than women on health and good looks (Table 5.1). In addition, a later study across forty-eight nations reported that, on average, men have more permissive sexual attitudes than women (Schmitt, 2005), supporting the idea that men and women differ in their attitudes towards promiscuous matings.

Women and men also showed similarities in their self-reported mate preferences, with the four most highly ranked characteristics being identical across the sexes (Table 5.1), and some cross-cultural variation was recorded (e.g. in the importance of 'chastity'). In 1999, Alice Eagly and Wendy Wood re-analysed Buss' dataset and reported that the extent to which women place

Table 5.1 Average ranking of eighteen characteristics in a potential partner by women and men

Ranking	Women	Men
1	Mutual attraction/love	Mutual attraction/love
2	Dependable character	Dependable character
3	Emotional stability/maturity	Emotional stability/maturity
4	Pleasing disposition	Pleasing disposition
5	Education/intelligence	Good health
6	Sociability	Education/intelligence
7	Good health	Sociability
8	Desire for home and children	Desire for home and children
9	Ambition/industrious	Refinement/neatness
10	Refinement/neatness	Good looks
11	Similar education	Ambition/industrious
12	Good financial prospects	Good cook and housekeeper
13	Good looks	Good financial prospects
14	Favourable social status	Similar education
15	Good cook and housekeeper	Favourable social status
16	Similar religious background	Chastity
17	Similar political background	Similar religious background
18	Chastity	Similar political background

Based on Table 4 in Buss (1990).

greater value than men on the financial resources of partners depended on the levels of gender equality in that particular country. Eagly and Wood reported that women and men ranked this trait more similarly in those countries with the highest levels of gender equality, which was measured by the extent to which both sexes participated equally in economic, political, and decision-making roles. These researchers argued that, in societies with higher levels of gender equality, women are less likely to be financially dependent upon their partners and therefore do not rank men's financial resources as being as important when entering into a relationship. Eagly and Wood (1999) instead proposed a *biosocial model* of sex differences, in which physical characteristics, such as the fact that women give birth and men have higher average muscle mass, make it more likely that certain gender roles, rather than others, will emerge within societies, with women characterized as homemakers and men as wage-earners. These gender stereotypes are then suggested to influence preferences, attitudes, and behaviour patterns, including mate preferences (Wood and Eagly, 2012).

Some Evolutionary Psychologists strongly refute the biosocial model, arguing that the majority of data on human mate preferences support the idea that men and women differ in their evolved psychological mechanisms (Gangestad and Simpson, 2007). Moreover, a large-scale replication of Buss' original study has provided evidence that the universal sex differences in mate preferences reported by Buss are robust (Walter et al., 2020). Some aspects of mate preferences do appear to vary systematically across cultures, sometimes in relation to gender equality indices. For instance, the higher the level of gender equality, the more likely it is that both men and women prefer partners of their own age (Walter et al., 2020). Evolutionary Psychologists have interpreted these types of findings as suggesting that variation in men and women's mating strategies reflects *conditional, context-dependent strategies*, such that universal evolved psychological mechanisms produce alternative strategies depending on an individual's specific local circumstances and personal characteristics (Gangestad and Simpson, 2000). Those individuals with the highest 'mate value' (i.e. who most closely match the ideal standard of the opposite sex) appear to have greater power of choice in mating markets and are most likely to have a partner who conforms to their ideal preferences (Conroy-Beam et al., 2019). However, the observed variation in mating strategies across individuals and across populations weakens the Evolutionary Psychology proposal that highly pre-specified cognitive mechanisms underpin mate preferences (Brown et al., 2009).

Violence and risk-taking behaviour

All around the world, folk literatures abound with 'Cinderella' stories that involve a cruel step-parent. For Martin Daly and Margo Wilson, two psychologists at McMaster University, Canada, the ubiquity of these stories reflected a genuine aspect of human societies. Daly and Wilson (1988) used an Evolutionary Psychology perspective to understand patterns of child homicide, predicting that step-parents will tend to feel less attached to children than biological parents, with the result that those individuals reared by people other than their biological parents will more often be at risk of neglect, physical harm, and death. In a study of infanticide in North America, Daly and Wilson documented a substantially elevated risk to children residing with one biological parent and one step-parent. For instance, out of 279 fatal incidents of child abuse detected by the American Humane Association in 1976, over 40% dwelt with a step-parent, which is considerably more than would be expected by chance. Although the vast majority of step-parents do not abuse or harm their step-children, several studies have provided supportive evidence that the rate of child abuse is somewhat higher in families with step-parents than in families with only biological parents (Archer, 2013).

An alternative explanation of these data is that the higher risks faced by step-children result from other variables that are also associated with being in a step-parent household. For instance, poverty is known to be a risk factor for child maltreatment, and step-parent households are more likely to occur in families in lower socioeconomic brackets. However, Daly and Wilson (2008) provided evidence that the risk from step-parents is an additive factor that is not fully explained by these potential confounds. Another potential explanatory factor would be if parents in step-families are more prone to exhibiting violent behaviour in general. Counter to this argument, higher rates of violence are reportedly shown to step-children than to biological children within individual families, which suggests that family environment does not fully explain the risk to step-children. Other critics have attempted to dismiss the 'Cinderella effect' as resulting from a reporting bias, on the assumption that welfare professionals take the presence of a step-parent to be diagnostic of whether a child's injury results from abuse or accident (Buller, 2005). Daly and Wilson (2005) argued that any such biases are unlikely to be large enough to account for the differences in abuse rates and suggested that the data instead confirm that parents discriminate in favour of their own offspring. Other Evolutionary Psychologists have taken the interpretation one step further, suggesting that homicide of costly, unrelated offspring might have been an adaptive strategy in past selective environments (Duntley and Buss, 2011).

Evolutionary Psychologists have also investigated murders of adults outside of the family. In the vast majority of human populations that have been studied, men are more likely than women to be the perpetrators, and the victims, of homicide. In order to explain this apparent universal sex difference in homicide rates, Daly and Wilson (1988) turned to sexual selection theory. In animals, aggressive behaviour can occur during competition for resources, which can include access to potential mates. In mammals, where females generally contribute more resources to raising offspring than do males, the competition for mates is predicted to be highest in males, assuming that access to mates is the factor that limits male reproductive success. In human beings, many male–male homicides result from seemingly trivial altercations over social status that escalate into violence, which has been proposed to reflect the importance of male status building in the small social groups that were inhabited by our ancestors (Liddle et al., 2012). If high-status males were likely to have the greatest mating success, the decision to engage in risky and violent behaviour might be worth taking, particularly in unpredictable environments where future opportunities might be limited. Consistent with this idea, the sex difference in mortality rates in the United States is highest for those individuals in the lowest income brackets who presumably face the most unstable futures (Kruger and Nesse, 2006). Violent conflicts between groups of men could also potentially be explained by selection having favoured male coalitionary and aggressive behaviour (McDonald et al., 2012).

In addition to exhibiting higher levels of violent behaviour, men engage in higher average levels of risk-taking and impulsive behaviour than women (Cross et al., 2011). These sex differences emerge during adolescence, when young men are particularly prone to dangerous driving, criminal behaviour, and drug misuse. Evolutionary Psychologists have suggested that these sex differences in response to risk again reflect the differential selection that acted on males and females in ancestral human populations (Ellis et al., 2012). Risk-taking potentially led to high rewards for males, in terms of access to status, resources, and mates, whereas similar behaviour in females arguably would have jeopardized their future reproductive success. Higher average levels of fear in women compared to men potentially underpin women's greater reluctance to engage in risky behaviour, including aggression (Cross and Campbell, 2011). Researchers have also proposed that absolute levels of risk-taking will vary according to the level of environmental unpredictability. In unpredictable or dangerous environments, selection is suggested to have favoured individuals who prioritize immediate benefits at the expense of longer-term planning. Cross-cultural data from seventy-seven countries have revealed that levels of risk-taking are higher for both men and women in harsh environments, compared to more benign environments, and that the drop in risk-taking with age is also less steep in such environments (Mata et al., 2016). These data are interpreted as implying that risk-taking is an evolved psychological mechanism that exhibits sex differences and is also sensitive to local environmental cues.

Disgust

Disgust is an emotion that is expressed in a diverse range of contexts. We may experience disgust at the idea of eating sheep's eyeballs or drinking sour milk, but we might also feel disgusted at the thought of sexual relations with someone twice our age or at a player who cheats in a football game. How can we explain the existence of these emotional responses? Darwin (1872) proposed that disgust is a universal human emotion and noted that it has an associated facial expression that is recognizable across widely different cultures. Evolutionary Psychologists have suggested that disgust is an adaptation that serves two functions, firstly to guard our bodies against disease and secondly to ostracize those individuals who break social contracts (Curtis and Biran, 2001). By motivating the avoidance of contaminated food and people who may be contagious, the experience of disgust is thought to have evolved because it reduced the likelihood that our ancestors would have eaten material containing pathogens and thereby reduced exposure to transmissible diseases. Those individuals who exhibited disgust and moral outrage when others violated social norms, and thus avoided contact with such individuals, might have benefited from instead engaging in safer, altruistic exchanges with individuals who followed through on social contracts. Evolutionary Psychologists have therefore proposed that disgust functions in both physical and moral domains.

Val Curtis and colleagues (2004), from the London School of Hygiene and Tropical Medicine, UK, devised several predictions based on the assumption that disgust partly arose to prevent disease transmission. For instance, disgust should be felt more strongly when an individual is faced with a disease-relevant stimulus than with a similar stimulus with no disease connotations; it should operate similarly across cultures; it should be more pronounced in women than men, as women play a double role in protecting both themselves and young offspring from disease; it should become less potent as an individual's reproductive potential declines; and it should be more strongly evoked by contact with strangers than close relatives, because strangers may carry novel pathogens. These predictions were tested using data provided by almost 40,000 participants in an international, online survey in which subjects were asked to grade the disgustingness of pairs of disease-relevant and non-relevant pictures across a variety of contexts. More than 98% of people found the disease-relevant pictures equally, or more, disgusting than the paired images, with the same patterns found across participants from virtually all cultural regions, and with the other predictions also confirmed. The hypothesis that disgust is an adaptation designed to prevent the acquisition of diseases was thus supported by the tests.

Although behavioural disease avoidance has obvious benefits, excessive avoidance measures might be costly, as rejecting potential food items could lead to greater time and energy spent foraging and a more restricted diet. Accordingly, Evolutionary Psychologists anticipate that disgust will be triggered in a highly specific manner, fashioned by selection to occur in situations where the benefits of food avoidance outweigh the costs of lost calories. Daniel Fessler and colleagues, at the University of California, Los Angeles, explored this hypothesis in a number of ways. They anticipated that elevated disgust would be triggered when an individual is particularly vulnerable to disease as a result of lowered functioning of the immune system. Changes in immune functioning over the course of pregnancy offered an opportunity to test this hypothesis (Fessler and Navarrete, 2003). During the first trimester of pregnancy, the mother's immune system is substantially suppressed, which prevents the mother's body from rejecting the developing foetus but increases the vulnerability of both the mother and foetus to pathogens. Food-borne illnesses are a particular threat. Using an online survey of pregnant women, the researchers confirmed their prediction that pregnant women in the first trimester would exhibit elevated disgust sensitivity, particularly in the food domain. These findings, which controlled for heightened nausea, supported the hypothesis that disgust sensitivity varies during pregnancy in a manner that compensates for vulnerability to disease.

Evolutionary accounts of disgust can be contrasted with an alternative perspective that views disgust as guarding the self against symbolic (rather than actual) pollution from the physical and social world (Rozin et al., 2000). However, Fessler and Haley (2006) argued that the empirical predictions generated by this hypothesis would also be consistent with an Evolutionary Psychology approach. The first step in their argument was that those parts of the body most associated with the self are those on the outside, which interface with the environment and are thereby most vulnerable to contamination. They then reasoned that, if disgust serves to shape behaviour so as to protect the body against pathogens and toxins, those parts of the body that interface with the environment (e.g. skin, eyes, and genitals) should be associated with disgust reactions to a greater extent than internal parts of the body (e.g. spleen, liver, and intestines). In an online study, participants were asked to imagine that it was the year 2050 and that medical science had progressed to the point where organ and tissue transplants were routine. They were then required to rank how disgusting it would be to receive a transplant for each of twenty parts of the body. As predicted, the transplanting of external body parts elicited stronger disgust responses than internal body parts. Thus, from an Evolutionary Psychology viewpoint, disgust responses have been shaped by past selection to be activated by specific elicitors in a manner that is consistent with a pathogen-avoidance function.

But if disgust evolved to protect the body from exposure to pathogens and toxins, why does it span such a variety of other behavioural domains? Evolutionary Psychologists have argued that the neurophysiological responses underlying disgust reactions to the threat of pathogen consumption were recruited by natural selection to generate adaptive behaviour in other domains (Curtis et al., 2004). For instance, Fessler and Navarrete (2003) proposed that sexual disgust is an adaptation that functions to inhibit participation in fitness-reducing sexual unions. Consistent with this argument, bestiality, incest, and other deviant sexual behaviours are considered disgusting cross-culturally, and, on average, women have a particularly high disgust sensitivity in the sexual domain relative to men (Tybur et al., 2009). However, other aspects of disgust, most notably those regarding which food items are edible, do vary across cultures, suggesting that any evolved mechanisms of disgust response interact with social learning processes to produce local traditions (Curtis et al., 2011). An important role for learning is also implied by a study reporting a reduced level of disgust in mothers compared to childless females, implying that disgust sensitivity may be suppressed by caring for children (Prokop and Fancovicova, 2016). Understanding how universal disgust reactions are expressed across different populations, different individuals, and the same individuals at different timepoints, and how these responses are modified by learned rules, has been used to enhance hygiene behaviour and improve public health.

Critical evaluation

Some of the criticism levelled at Evolutionary Psychology is identical to that directed at sociobiology; indeed, many critics see no meaningful distinction between these two schools (e.g. Rose and Rose, 2000). Rather than repeat ourselves, we refer the reader back to the penultimate section of Chapter 3, where we discuss these charges. To reiterate briefly, we argue that allegations of genetic determinism or prejudice on the part of leading sociobiologists or Evolutionary Psychologists are usually unfounded. However, Evolutionary Psychology, like human sociobiology, is vulnerable to the charge of 'just-so' storytelling. Even the most avid Evolutionary Psychologist would have to admit that their field has produced its fair share of weak studies and unsupported narratives. Rather than pointing to such studies individually, here we attempt to provide a fair and balanced evaluation of the strengths and weaknesses of the theoretical foundations of the sub-field. This balance is important because there seemingly is, as writer Robert Wright characterizes it, an 'anti-evolutionary psychology market niche' (cited in Daly and Wilson, 2008, p. 396), where hostile detractors pour scorn on a straw-man caricature of the sub-field. In our judgement, uninformed and unfair denigration is just as counterproductive as poor research. In the following sections, we evaluate issues and points of contention related to the EEA, the Evolutionary Psychologists' general evolutionary perspective, and the emphasis on domain specificity.

Evaluating the concept of the environment of evolutionary adaptedness

Much of the dissatisfaction with the EEA concept has derived from the stereotyped characterization of our ancestors as rudimentary hunters who roamed ancient African grasslands with stone tools in hand (Foley, 1995). Cosmides and Tooby (1987) made reference to the observation that 'humans spent 99 percent of their evolutionary history as hunter–gatherers in Pleistocene environments' (p. 280). Yet Cosmides and Tooby have informed us that they never adhered to a stereotyped view of the EEA and that their early writings on the EEA were simplified to reach an 'evolutionarily naïve' audience or those that tended to regard all human behaviour as being of utility in current environments. Unfortunately, a damaging EEA-as-Pleistocene-African-savannah stereotype pervaded the early Evolutionary Psychology literature and is often an unspoken assumption within current research.

What is wrong with the notion of the human EEA as a particular time and place? One problem is that the Pleistocene is a very long time period that encompasses the evolutionary history of numerous ancestral hominins, which raises the question of which ancestors we imagine are represented in the EEA. At the start of the Pleistocene (around 2.8 million years BP), the earliest species within the *Homo* genus emerge in the fossil record (e.g. *Homo habilis* and *Homo ergaster*). These early *Homo* species had larger brains and smaller teeth than their ape-like ancestors and made simple stone tools, but depictions of their behaviour, social lives, and cognitive abilities remain largely speculative. Our own species (*Homo sapiens*) is thought to have diverged from our closest extinct relative, *Homo neanderthalensis*, around half a million years ago (i.e. late Pleistocene). Many human behavioural and cognitive traits must have been selected prior to the emergence of anatomically modern *Homo sapiens*, yet comparatively little is known about the lifestyle of our ancestors through the early and mid-Pleistocene period (e.g. whether they had sophisticated linguistic abilities, shared food, and had home bases). The hominin fossil record is limited in terms of both the amount and distribution of evidence, which poses challenges for asking even relatively basic questions about hominin evolution, such as whether specific climatic and environmental parameters might have influenced brain size expansion across the lineage (Faith et al., 2021).

A stereotypical notion of the EEA as an African savannah also implies that Pleistocene hunter–gatherers exhibited little variability in space, which evidence shows is false. During the early Pleistocene, our hominin ancestors lived not only in grasslands, but next to rivers and oceans, and in wetlands, woodlands, and both temperate and tropical forests (Foister et al., 2023; Zeller et al., 2023). These early *Homo* species also lived outside of Africa, including in Europe, the Asian mainland, and south-east Asian islands. The fossil record reveals several subsequent waves of expansion of *Homo sapiens* out of Africa, including a worldwide expansion that occurred around 40,000–50,000 years BP (Bergström et al., 2021). Describing our ancestors as 'hunter–gatherers' is also an insufficient account of their life histories. Modern hunter–gatherer populations exhibit substantial diversity in their subsistence patterns and life histories both between and within populations (Kelly, 2013; Page and French, 2020), and ancestral hominin populations that spanned substantial time periods and geographic ranges are likely to have exhibited extensive diversity. Many archaeologists and anthropologists believe that ancestral *Homo* species, and even early *Homo sapiens*, lived completely different lives to modern hunter–gatherers. Finally, as discussed below and in Chapter 7, the evolution of our own species did not finish at the end of the Pleistocene and continues to the present day.

Given these challenges, can the concept of the EEA be put to use, in the manner that Tooby and Cosmides (1989a) originally claimed, to develop models of adaptive problems the human psyche had to solve? If the EEA is defined as 'a statistical composite of the adaptation relevant properties of ancestral environments' (Tooby and Cosmides, 1990a, pp. 386–387), can researchers realistically compute a statistical composite of all the relevant environments encountered by our ancestors, and weight them accordingly? Tooby and Cosmides' (2005) description of ancestral hominins as long-lived, ground-living primates with an omnivorous diet, altricial young, male–female bonds, and a sexual division of labour perhaps does not provide sufficient detail to formulate specific predictions. If researchers are going to use the EEA as originally outlined, they presumably need to identify a particular time period and group of ancestors when the relevant psychological mechanisms evolved, and then weight that and all subsequent environments accordingly. In principle, EEA supporters could carry out a phylogenetic analysis of the traits of ancestral humans to determine the earliest known ancestor that exhibited a particular trait, which would establish when the trait of interest first appeared. In practice, this approach is rarely taken and, as little is likely to be known about that particular ancestor and most of its descendants, it would be a time-consuming exercise that would generate only vague speculation.

The EEA concept has engendered a wealth of undisciplined speculation and storytelling in which virtually any attribute can be regarded as an adaptation to a bygone world. The real virtue of the EEA concept is perhaps more modest. The EEA idea encourages researchers to recognize that humans, like all species, exhibit *some* adaptations to past environments that are not necessarily of current utility. The originator of the EEA concept, John Bowlby, was concerned with the mother–child relationship, which we might envisage has a degree of constancy across environments and over time. There is a strong argument that the EEA concept was important in developing an understanding of childhood separation anxiety and attachment (Hinde, 1987). Similarly, researchers do not need to know the precise conditions in which humans evolved to make the reasonable guess that salts and sugars may not have been in abundant supply so that their reinforcing properties may not have been counterbalanced by regulatory processes operating against consuming excess (Bateson and Martin, 1999). However, the lack of detailed knowledge about ancestral hominin populations is likely to restrict the type of question that can be asked based on robust assumptions about past selective environments.

Advances in evolutionary biology

Even if researchers could reliably determine which traits our Pleistocene ancestors had, couldn't humans have evolved psychological mechanisms that have been shaped by selection prior to, or after, the Pleistocene? In conversation with us, Tooby suggested that selection prior to the Pleistocene is unlikely to be important because selection during the Pleistocene would have refashioned human psychological adaptations to be suited to Pleistocene conditions. In terms of more recent evolution, Cosmides and Tooby have argued that:

> the complex architecture of the human psyche can be expected to have assumed approximately modern form during the Pleistocene ... and to have undergone only minor modifications since then. (1987, p. 34)

This reasoning, which is consistent with the idea of *gradualism*, is based on the assumption that evolutionary change occurs at a slow rate; yet it is far from clear that this assumption, nor the assumption that there has been no meaningful selection on psychological mechanisms since the end of the Pleistocene, is correct. These assumptions were not unreasonable in the 1980s and 1990s, but they are now challenged by evidence showing more rapid rates of biological evolution in animals than hitherto anticipated (Bonnet et al., 2022; Kingsolver et al., 2001). In recent years, evidence has been found for natural selection acting on contemporary human populations (Stearns et al., 2010), and geneticists have also uncovered evidence that several hundred genes have been subject to recent selection in human beings

(López Herráez et al., 2009; Voight et al., 2006; Wang et al., 2006; Williamson et al., 2007). In Chapter 7, we discuss several human genes expressed in the brain that fall into this category. Indeed, there is evidence that the human brain actually evolved more rapidly than other human tissues in recent millennia (Khaitovich et al., 2006).

These advances in our understanding of genetic evolution imply that there could have been significant selection of human psychological traits since the end of the Pleistocene (Bolhuis et al., 2011). In addition, the suggestion that Evolutionary Psychologists believe human beings are exclusively adapted to a past world and not at all adapted to modern life is said to be a 'straw man' (Barrett, 2015). In the Holocene (the epoch following the Pleistocene), human populations have exploded, and human beings have colonized the globe. This population growth and expansion suggests that a significant fraction of human characteristics remain adaptive even in modern environments (Laland and Brown, 2006). While our evolved psychological mechanisms have provided us with the cognitive machinery to cope with, and in fact produce, these changes in our environment, any assumption that natural selection on humans has stopped, or that no genetic variation underlies human psychological characters, is no longer tenable. Hence, there is little reason to assume that our evolved psychological mechanisms could not have been subject to recent selection or, more generally, that our cognition could not adjust to contemporary worlds.

Moreover, the view that modern human populations are primarily adapted to an ancestral Pleistocene habitat is misleading in another respect: it portrays humans as passive victims of selection rather than as potent constructors of their niche. It is a distortion to regard evolution as a process by which organisms solve problems set by the environment (Lewontin, 1983). Niche-construction theory lays emphasis on the fact that organisms themselves *modify* important components of their selective environments with evolutionary consequences (see Chapter 7; Odling-Smee et al., 2003), a point now accepted by some Evolutionary Psychologists (Barrett and Armstrong, 2023). For human beings, our capacity to continuously create solutions to self-imposed problems reflects the fact that we are very adaptable creatures. To a degree that surpasses other species, human mental processes must contend with a constantly changing information environment of their own creation (Laland, 2017). Evolutionary Psychologists have stressed that our evolved psychological mechanisms do not necessarily produce adaptive responses in twentieth-century industrialized environments (Cosmides and Tooby, 1987). In contrast, the niche-construction perspective anticipates that humans will experience less 'mismatch', or 'adaptive lag', than other animals, given that, when humans modify their world, they are likely to do so in a manner that is guided by, and suited to, their adaptations.

The fact that few Evolutionary Psychology studies refer to the findings of modern evolutionary biology reinforces the concern that Evolutionary Psychology has become detached from recent developments in evolutionary thinking, which over the last forty years have increasingly stressed a wide range of processes. John Endler (1986b) identified at least twenty-one processes that are instrumental in evolutionary change. Only one of these processes is natural selection, others being drift, mutation, and so forth. It has also become clear that natural selection operates at several different levels and, unlike forty years ago, multi-level selection models are now a common and respectable feature of evolutionary genetics. Selfish DNA such as microsatellites, and selfish genes such as transposons and segregation distorters, are examples of selective processes operating below the level of the individual, while above this level an increasing proportion of evolutionary biologists accept the idea that species selection and clade selection could be important (Eldredge et al., 2016). Group selection has increasingly been put forward as an important process in human evolution, as discussed in Chapter 7. Evolutionary Psychology has perhaps been more influenced by popular evolution writings, as exemplified by the books of Richard Dawkins, and by superficial interpretations of the work of Darwin, Trivers, and others, rather than the more technical primary evolutionary biology literature. The reality is that evolution is a far more complex phenomenon than generally portrayed in Evolutionary Psychology textbooks.

Most Evolutionary Psychologists adhere to a branch of evolutionary thinking known as *adaptationism*. Unfortunately, this term is used in multiple ways both by enthusiasts and critics of this perspective, which can lead to further disagreements. Some adaptationists take inspiration from George Williams' (1966) book *Adaptation and Natural Selection*, which advocated for a more rigorous use of the term 'adaptation' than previously advanced and argued that natural selection was a sufficient process to explain most of what is important about evolution. Despite this call for greater rigour, for their critics, adaptationists are researchers who describe virtually all characters as adaptations and who underestimate the importance of other processes in evolution. As an example of adaptationism, it is well known that human evolution is characterized by 'neoteny', that is, a slowing down in development, so that, for certain characteristics, the anatomy of an adult human resembles an infant ape more than it resembles an adult ape. Lewontin (2000) pointed out that, while there have been many speculations about why natural selection might have favoured a protruding chin in humans, this character may actually be a by-product of selection on other features (see definition in Figure 4.1 in Chapter 4). Changes in human diet and cooking have resulted in reduced dentition and retraction of the lower and upper jaw at different rates, and the chin may be an incidental outcome. In other words, the chin is probably not a character that has been favoured by natural selection (although this issue remains to be fully resolved; Pampush and Daegling, 2016).

Identifying what constitutes a character that is subject to natural selection is a well-recognized and stubborn problem within contemporary evolutionary biology, which has countless difficulties but no universally accepted solution (Wagner, 2001). While many Evolutionary Psychologists are commendably disciplined in their attribution of adaptations, which are carefully distinguished from exaptations and by-products (for definitions, see Chapter 4), others appear less cautious. Critics of adaptationism feel that these researchers underestimate the significance of evolutionary processes other than the natural selection (Lloyd and Feldman, 2002; Lynch, 2007). If evolution is a complex multifaceted phenomenon, if many evolutionary processes, including drift and mutation, are operating at the same time, if evolutionary history is important, if selection is operating at different levels, if evolutionary rates can sometimes be fast, and if evolutionary theory is rapidly developing, it makes the business of predicting and interpreting psychological adaptations much more difficult. However, there is no virtue in pretending that evolution is a simpler process than it actually is. In addition, a key point is that Evolutionary Psychologists commonly fail to conduct studies that test between competing evolutionary explanations, such as those derived from cultural evolution theory (described in Chapter 6), and non-evolutionary explanations (Bolhuis et al., 2011). Creating an evolutionary 'just-so' story that is perhaps consistent with a piece of empirical data is not sufficient to provide supportive evidence for that specific evolutionary scenario; researchers must compare their evolutionary explanation with alternative evolutionary, and non-evolutionary, explanations, as their data might be consistent with more than one hypothesis. Ideally, researchers would generate predictions that would help them to distinguish between a range of alternative explanations.

Psychological traits may be domain-general

One contentious aspect of Evolutionary Psychology is the stress laid on domain-specific psychological modules. Cosmides and Tooby (1987) characterized the difference between their perspective and the standard social science view as representing a choice between two models of the mind, one that lays emphasis on a large number of domain-specific modules versus another stressing a small number of domain-general processes. Many researchers believe that Evolutionary Psychologists have overplayed the modularity of the human brain, and maintain that minds have many domain-general features (Roberts, 2007). A number of Evolutionary Psychologists have presented the case for domain-general evolved psychological mechanisms (Atkinson and Wheeler, 2004; Fessler and Machery, 2012). For example, human intelligence may be a domain-general trait that, by definition, functions across multiple domains and that provides fitness-relevant skills, such as the ability to innovate and use social information, across a range of social and ecological contexts (Burkart et al., 2017). However, in spite of these developments, a modularized depiction of the mind continues to dominate the Evolutionary Psychology literature, and Evolutionary Psychologists continue to give short shrift to the various roles that general rules, processes, and regulatory mechanisms might play.

However, domain-specific and domain-general represent poles of a continuum in descriptions of how the mind works. Evolutionary Psychologists are surely correct to point out that there are efficiency benefits to be gained by mental division of labour and that, at times, evolution would favour specialization of psychological processing. Yet one can also have too much specificity, as early students of artificial intelligence established (McCorduck, 2004). To take the idea of domain specificity to its logical extreme, in principle, it might be possible to construct a central nervous system that contains specific modules that individually respond to any possible domain-relevant event in the environment, but the cost in terms of neural circuitry would be huge. Evolutionary Psychologists do not go to this extreme, but they do envisage the existence of a large number of evolved mechanisms in the brain, each addressing a class of adaptive problems. Some Evolutionary Psychologists suggest that domain-general cognitive mechanisms are unlikely to evolve (Buss, 2008) or that 'every mental adaptation is specialized' (Barrett, 2015, p. 26). However, general solutions may be favoured when they can do a good enough job at low cost, and domain-general processes are no more incompatible with evolutionary theory than are domain-specific processes.

The experiments on learning predispositions in rats and monkeys that were described earlier are frequently hailed by Evolutionary Psychologists as demonstrating species-specific constraints on learning. Yet associative learning is widespread and has general properties that allow animals to learn about the causal relationships among a wide variety of events (van Duijn, 2017). Learning can therefore occur via quite simple rules; for example, one theory, known as the Rescorla–Wagner rule, has proved useful in explaining the results of experiments on foraging in honeybees, avoidance conditioning in goldfish, and inferential reasoning in humans. Even some of the most enthusiastic supporters of a modular view of the brain (e.g. Shettleworth, 2000) accept that, while what is learned may vary adaptively across species, *how* it is learned does not. Natural selection may have fashioned us to be prepared to form some associations more readily than others, and built in some motivational priorities, but many psychologists regard this specialization as more tinkering with the general system than constructing an independent set of species-specific learning processes (Bolhuis and MacPhail, 2001).

Cosmides and Tooby (1987) have argued that learning should not be regarded as an alternative to evolutionary explanations. However, our capacity to learn is an unusual adaptation. It has a property that makes it different from many other adaptive responses of phenotypes to the environment, such as calluses on the hands; namely, it is an information-gaining subsystem. The function of learning is to acquire and store information about the world, information that will generally guide behaviour towards adaptive goals, but information that nonetheless could not be specified in our genes. Rather than fashioning us with brains hardwired to recognize apples as food and sand as not food, natural selection has given us a flexible information-gaining problem solver, with instructions to seek food when blood sugar levels are low and to recognize apples as food because they taste good whereas sand does not, as discovered by numerous infants on their first trip to a beach. A rule like 'Actions that are followed by a positive outcome are likely to be repeated, while those followed by a negative outcome will be eliminated' is domain-general in the sense that it can be equally applied to behaviour concerned with finding food, avoiding predators, or seeking a mate. This particular rule, which was first described by American psychologist Edward Thorndike in 1911 and is known as the 'Law of Effect', is likely to govern much human learning.

Brain development operates on similar principles. The developing nervous system generates excess neurons and excess neuronal connections, sending out neural projections at random, then prunes these to retain the most effective. The final anatomy depends very much on experience. For example, neuroscience experiments reveal that cats raised in a world of horizontal lines are 'blind' to vertical objects (Hubel and Weisel, 2005). Much of the debate over the merits of Evolutionary Psychology explanations revolves around the extent to which human developmental processes are under tight genetic regulation in which developmental outcomes are pre-specified and channelled (e.g. via evolved reaction norms; Barrett, 2015), as opposed to more flexible systems in which pre-specification of regulatory development is more modest (Karmiloff-Smith, 2000). Given the immense developmental plasticity and flexibility of the human brain, it is conceivable that much human cognition and behaviour is better regarded as resulting from our extraordinary adaptability.

Few parts of the human brain, or cognitive domains, are strictly modular in an encapsulated way, where encapsulation is defined as a tightly restricted flow of information into the neural pathways or functional unit. Extensive evidence has been found for top-down effects, cross-domain integration, and cross-modal neural plasticity, and overlapping neural circuitry can perform functionally distinct tasks (Anderson and Finlay, 2014). A meta-analysis of data from numerous brain-scanning studies established that, independent of scale, a typical human brain region is activated by tasks in multiple different domains (Anderson, 2010). Broca's area, for instance, is a region of the brain that is strongly associated with language but is also involved in movement preparation, action sequencing, action recognition, and motor imitation. The widespread reuse of neural circuitry across different domains in the human brain is incompatible with both the idea of structural modularity and with encapsulated conceptions of functional modularity.

It is even conceivable that cognitive modularity has been reduced during recent human evolution, allowing more integration of information and communication among modules (Mithen, 1996). Traits commonly evolve by reusing pre-existing structures (from gene regulatory networks to neural circuitry) and, consistent with this observation, neuroscientific data suggest that more recently evolved functions of the human brain exploit a greater number of widely scattered brain areas than do evolutionarily older functions (Anderson, 2010). For example, language and reasoning are recently evolved capabilities and exhibit scattered circuitry that integrates spatially separate regions of the brain. Plausibly, extensive reuse of neural circuitry has generated greater neural integration in humans than most other animals. In summary, a version of 'massive modularity' that regards cognition as modular right through from perception to action, with modules rarely interacting, is unlikely to reflect how the brain works (Anderson, 2010; Fodor, 2000). More recent Evolutionary Psychology accounts, while retaining an emphasis on domain specificity, acknowledge this interaction between domains (Barrett, 2015).

Conclusion

There are undoubtedly some fine pieces of Evolutionary Psychology work that show genuine promise of being able to decipher the evolved structures of the human mind. The best of Evolutionary Psychology is as rigorous and sophisticated as any research carried out in the general area of human behaviour and evolution. At the same time, the discipline is marred by a number of weak studies that do little more than use a Pleistocene stereotype to contrive a 'just-so' evolutionary story or that assume the universality of a trait without considering the role of developmental and social processes. Sadly, these poorer studies frequently have a sensational quality that results in their receiving considerable public attention, particularly when such studies bolster stereotyped and prejudiced world views. Perhaps too much research in the field is a documentation of what is already observed, accompanied by a post hoc evolutionary spin and a snappy press release. It would be unfair to condemn the entire field of Evolutionary Psychology on the basis of the work of its weakest practitioners. However, numerous psychologists, biologists, anthropologists, and philosophers have stressed the need for more sophisticated approaches to psychological phenomena than are typical of Evolutionary Psychology and have called for greater integration with adjacent fields, such as cognitive science and neuroscience (Barrett, 2024; Barrett et al., 2014; Bolhuis et al., 2011; Heyes, 2000; Lickliter and Honeycutt, 2013; Muthukrishna and Henrich, 2019; Narvaez et al., 2022; Stotz, 2014).

We began this chapter by emphasizing the historical importance of the Santa Barbara school, and we now return to discuss to what extent it remains representative of the field as a whole. Certainly, we do not believe that our portrayal of Evolutionary Psychology is entirely anachronistic, and we retain the view that the bulk of the central tenets of the Santa Barbara school characterize the broader sub-field. In our judgement, the philosophical, conceptual, and methodological structures developed by the leaders of Evolutionary Psychology in the 1980s remain the dominant influence on the field, and universalism, gradualism, domain specificity, gene-centrism, the EEA, and so forth, remain widespread. These ideas are, for instance, prominent in leading contemporary Evolutionary Psychology textbooks and edited volumes (e.g. Buss, 2015, 2019; Shackelford, 2020). Nonetheless, we see positive signs, for instance, of some integration between Evolutionary Psychology and Human Behavioural Ecology perspectives, of a greater wariness of EEA storytelling, and of individual researchers utilizing methods from across sub-disciplines. We believe that Evolutionary Psychology can be a progressive research programme, if debates about the relevance of advances in evolutionary biology to psychology could be resolved (Bjorklund, 2016; Witherington and Lickliter, 2016), as discussed further in Chapter 8.

There is one criticism of Evolutionary Psychology on which we have not yet dwelt, namely that it underestimates the critical role of cultural transmission processes in shaping human knowledge and behaviour. In the next two chapters, Chapters 6 and 7, we will consider evolutionary perspectives that treat culture as a much more dynamic and influential process than hitherto regarded. Maybe social scientists are right to view cultural processes as not always well specified by our genes or environment, and as having a limited autonomy from biological control. Perhaps culture is an important evolutionary player in its own right.

6
Cultural evolution

Daniel Dennett describes Darwin's theory of natural selection as like a universal acid that 'eats through just about every traditional concept, and leaves in its wake a revolutionised world-view' (1995, p. 63). As an explanatory abstraction, natural selection is seemingly too good an idea to be restricted to biological evolution. As soon as *The Origin of Species* was published, scientists and philosophers began to speculate as to whether other entities, such as the immune system or the central nervous system, might also change and adapt through a selective process, while social scientists suggested that even scientific theories themselves undergo evolutionary change. Following previous scholars (van Wyhe, 2005), Darwin suggested that languages are evolving, stating boldly in *The Descent of Man* that, 'The survival or preservation of certain favoured words in the struggle for existence *is* natural selection' (1871, p. 60–61, italics added).

Darwin's intuition that natural selection may be a general law for how a multitude of processes change has proven plausible. For instance, the adaptive immune system generates antibodies and T cells with initially random variation, and then internal selection multiplies and refines those that bind successfully to antigens, with a memory of effective molecules retained (Klenerman, 2017). The vascular and nervous systems, animal learning, and much collective animal behaviour all operate on similar selective principles. There is a respectable scientific and philosophical tradition, somewhat esoterically known as *evolutionary epistemology*, that stresses the universal nature of natural selection (Plotkin, 1982, 1994) and that is backed by luminary philosophers (Dennett, 1995; Hull, 1982; Popper, 1979) and Nobel Prize-winning scientists (Edelman, 1987; Lorenz, 1965). Henry Plotkin's book, *Darwin Machines and the Nature of Knowledge* (1994), is a compelling exposition of this perspective. Plotkin and others attempt to see which phenomena, in addition to selection on genes, can be fruitfully treated as selection processes.

While Darwinism continues to 'eat' its way voraciously through countless academic disciplines, the social sciences stand out as a last hold of resistance. Stalwarts of the humanities have for many years maintained that no biological theory is going to explain much about human cultural change. A great deal of what is interesting about humanity, it is claimed, cannot be explained in terms of genes or fitness. Explanations in terms of culture are, for most social scientists, more compelling than biological accounts. What then if culture itself evolves? The idea that culture evolves predates Darwin's writings, and a linear, progressive conception of societal evolution was championed by some late nineteenth- and early twentieth-century anthropologists, and still has its advocates today (Caneiro, 2003). However, the development of a genuinely Darwinian theory of cultural evolution remained in its infancy throughout most of the twentieth century until, prompted by the human sociobiology debate, the idea that cultural practices change over time in a broadly Darwinian manner began to come back into favour. In the 1970s, the modern field of Cultural Evolution surfaced (henceforth, this phrase will be capitalized to denote this specific sub-field, whereas the phenomenon of 'cultural evolution' will be denoted using lower-case letters), deriving impetus from two theoretical developments. One was the emergence of *memetics*.

In the final chapter of *The Selfish Gene* (1976), Richard Dawkins let loose a new *cultural* replicator. Stressing the similarity between cultural and genetic transmission, Dawkins suggested that fashions, diets, customs, language, art, and technology evolve over historical time. He used the terms 'replicator' and 'vehicle' to distinguish between the 'immortal' genes, which are replicated in each generation, and the transient, vehicular organisms that house them. The gene is the archetypal replicator, but Dawkins proposed that a new, frequently insidious, kind of replicator has recently emerged on this planet, a mind virus that infects human beings with catchy concepts and fashionable ideas. He labelled this new replicator the 'meme'. Dawkins described how, just as soon as genes had blessed this particular species of 'lumbering robots' (1976, p. 19) with an enhanced capacity for imitation, memes set in. There was, as Dennett puts it, 'an invasion of the body snatchers' (1995, p. 342). Memes were described as having parasitized our vulnerable human brains, turning them into vehicles for their own virulent propagation.

According to Dawkins (1976), memes possess variation, heredity, and differential fitness, the three characteristics that are necessary for evolution by natural selection. They also commonly exhibit the qualities of particularly effective replicators—*longevity* (they frequently stay in our heads for long periods), *fecundity* (they can be copied and spread rapidly), and *copying fidelity* (at least some core components of memes are reasonably faithfully reproduced). In the rich environment of human minds, these characteristics may be all that memes need in order to evolve. According to Dawkins, 'memes propagate themselves in the meme pool by leaping from brain to brain via a process which, in the broad sense, can be called imitation' (1976, p. 206). He suggested that we don't pick or choose our ideas and beliefs; on the contrary, they pick and choose us, and manipulate us to their own ends. For Dawkins, meme evolution is not merely a process that can be metaphorically described in evolutionary terms; it *is* evolution by natural selection.

As scientific concepts go, the 'meme' meme had the best possible start: it was launched in one of the most popular scientific books of the twentieth century and given further attention through the highly successful writings of philosopher Daniel Dennett. In *Consciousness Explained* (1991), Dennett made memes the centrepiece of a grand theory of the evolution of mind. Dennett argued, somewhat disturbingly, that the human mind is an artefact created *by* memes *for* memes. Four years later, Dennett produced *Darwin's Dangerous Idea* (1995), another bestseller, this time a more general advocation of universal Darwinism, and again with memes as a central concept. Further popular books followed, including Aaron Lynch's (1996) *Thought Contagion: How Belief Spreads through Society,* Richard Brodie's (1996) *Virus of the Mind: The New Science of the Meme,* and Susan Blackmore's (1999) *The Meme Machine.* In the late 1990s, the *Journal of Memetics* was founded, and the first academic conferences on the topic were held, raising the possibility that memetics might emerge as an active research programme. Yet, the meme concept failed to take off as a serious explanation for cultural phenomena. The *Journal of Memetics* closed down in 2005, and interest in memes, at least in the scientific forum, waned. The reasons for this decline are multiple, but in no small part relate to the failure of meme enthusiasts to deliver a rigorous empirical or theoretical research programme.

In spite of the demise of memetics, the broader idea of cultural evolution had already spawned a rigorous science that spans biology, psychology, and anthropology. The emerging field built primarily on a second theoretical development of the 1970s—the mathematical modelling of cultural change. Like meme enthusiasts, the researchers who devised this approach argued that culture can be conceptualized as consisting of variants that compete with each other, in a similar way to competing alleles or genotypes. Unlike memetics, however, these researchers studied cultural change using theoretical models and methods adapted from contemporary evolutionary theory. Cultural evolution modelling builds on the strong theoretical traditions of evolutionary biology, but tailors its mathematical models and methods to the specific and unique processes of culture. This approach does not lead to the long-discredited conclusion that some societies are more advanced or superior to others (see Chapter 2), but it does suggest that some cultural traits are more likely to spread successfully than others, and it aims to explain and predict this pattern of change and diversity by identifying the underlying processes.

Quantitative study of cultural evolution began three years prior to the publication of Dawkins' *The Selfish Gene*. In 1973, two of the world's leading human geneticists, Luca Cavalli-Sforza and Marcus Feldman at Stanford University, California, USA, published the first mathematical models of cultural inheritance. Along with a substantial number of co-workers, Feldman and Cavalli-Sforza built up an impressive body of mathematical theory exploring the processes of cultural change. They frequently took advantage of the parallels between the spread of a gene and the diffusion of a cultural innovation to borrow or adapt established models from population genetics. Feldman and Cavalli-Sforza laid the theoretical foundations for the new research field of Cultural Evolution. Other mathematically minded researchers joined this new sub-field, most notably University of California anthropologists Robert Boyd and Peter Richerson, whose *Culture and the Evolutionary Process* (1985) introduced a variety of novel theoretical methods and ideas. Further impetus came from the application of phylogenetic methods to interpret aspects of human cultural variation, from languages to marriage systems. While the discipline initially prospered almost entirely through mathematical approaches, researchers gradually started to generate experiments and empirical studies, and, in 2016, the *Cultural Evolution Society* was founded. There now exists a vibrant sub-field of Cultural Evolution that has expanded to cover many current topics, including environmental sustainability, social media, the arts, and religion (Tehrani et al., 2023), and that has encouraged anthropologists, archaeologists, economists, historians, political scientists, and psychologists to ask questions about how culture evolves (Mesoudi, 2011).

Key concepts

In this section, we take a closer look at some key concepts associated with the sub-field of Cultural Evolution. We begin by asking how 'culture' has been defined, then describe some parallels between biological and cultural evolution, and finally examine different types of cultural selection.

Definition of culture

Historically, anthropologists have characterized culture as 'that complex whole which includes knowledge, belief, art, morals, custom and any other capabilities and habits acquired by man as a member of society' (Tylor, 1871, p. 1). However, this amorphous definition is not particularly conducive to scientific analysis. Human culture has proven a difficult concept to pin down, and there exists little definitional consensus within the social sciences, and frequently little appetite for definition (Durham, 1991). In this vacuum, Cultural Evolutionists, eager to explore how aspects of culture change over time, have taken a pragmatic line to operationalize culture. For most of this community, 'culture is information capable of affecting individuals' behaviour that they acquire from members of their species through teaching, imitation, and other forms of social transmission' (Richerson and Boyd, 2005, p. 5). Here 'information' includes knowledge, beliefs, values, and attitudes.

Once culture is defined as consisting of packages or distributions of socially learned information, the processes by which different variants of this information change in frequency in a population can be studied. The evolution of culture is characterized as a Darwinian process comprising the selective retention of favourable socially transmitted variants, as well as a variety of non-selective processes, including drift, migration, and invention (Boyd and Richerson, 1985; Cavalli-Sforza and Feldman, 1981). Rather than attempting to describe the entire culture of a society, culture is broken down into specific traits or variants (e.g. whether or not a particular foodstuff is eaten, or different re-tellings of the story of Little Red Riding Hood), allowing their frequencies to be tracked mathematically (e.g. Tehrani, 2013). This approach means that specific aspects of culture can be investigated in a straightforward manner. This broad characterization does not capture some features of culture that many social anthropologists deem to be important, such as shared intentions and values, ethnic boundaries, and moral codes. However, for Cultural Evolutionists, what this definition lacks in nuance, it makes up for in practicality, as it allows simple models to be formulated and tested. These models can then be, and indeed have been, applied to explore the evolution of socially constructed features, such as ethnic markers and shared norms (Richerson and Boyd, 2005).

A broad characterization of culture also opens up the possibility of 'culture' in non-human animals. Traditions for exploiting prey or food sites, tool-use, and vocalizations have been reported in a variety of animal species, including fish, birds, cetaceans, and non-human primates (Aplin, 2019; Laland and Galef, 2009; Whiten et al., 2017). The best-known case is the distinctive tool-using traditions of different populations of chimpanzees (*Pan troglodytes*) throughout Africa (Whiten et al., 1999). By comparing the behaviour of chimpanzees at seven different study sites across Africa, primatologists found that forty-two out of sixty-five recorded categories of behaviour exhibited significant variability across sites. For example, at some sites, chimpanzees used stone hammers to pound open nuts, a behaviour that was not observed at other locations, while the exact tool of choice sometimes varied between populations, such as use of either a long stick or short stick to dip for ants. Adoption of a broad notion of culture, one that promotes the investigation of social learning and traditions in other animals, has helped to shed light on the evolutionary roots of the unique human cultural capability, allowing for many cross-species comparative insights into human cognition (Hoppitt et al., 2008; Laland and Seed, 2021; Whiten et al., 2017).

Traditions and culture exhibit several properties of interest to evolutionary biologists. Perhaps the most obvious is that they are a source of locally adaptive behaviour; individuals can efficiently acquire solutions to problems such as 'what to eat' and 'with whom to mate' by copying others. But a variety of studies, ranging from investigations of human foraging traditions to fish mating sites, have established a capability of culture to propagate behaviour in a manner that is to some degree independent of the ecological environment. For example, Guglielmino and colleagues (1995) carried out an analysis of human cultural variation among 277 contemporary African societies and found that most traits examined correlated with cultural history rather than with ecology. Likewise, an analysis of behavioural variation in 172 small-scale North American societies found that cultural history had a stronger effect than ecology for the majority of behavioural traits (Mathew and Perreault, 2015). Such findings suggest that much of the between-population variation in human behaviour is maintained as cultural traditions that are distinct from ecological variables, although other aspects of human cultural variation may be evoked by the natural environment (Wormley et al., 2023). Genes and environment undoubtedly account for some variation in human behaviour, but the socially transmitted component of culture is impossible to ignore.

Parallels between biological and cultural evolution

In *The Origin of Species*, Darwin (1859) painstakingly presented extensive evidence for *variation* in the characteristics of individuals within a species, for *competition* among individuals for survival and reproduction, and for the *inheritance* of characteristics to the next generation. Culture equally exhibits these three characteristics (Mesoudi et al., 2004). Human culture clearly is extremely *variable*. Evidence for the diversity of cultural knowledge and artefacts is illustrated by the 16 million patents issued in the United States since 1963. The number of 'culture elements' (e.g. pottery, bows, shamanism, polyandry) in various Native American groups has been estimated to be 3,000–6,000, while the United States military force that landed in Casablanca during the Second World War was reportedly equipped with over 500,000 different material items (Steward, 1955). Similarly, Karl Marx famously expressed surprise at learning that, in 1867, over 500 different types of hammers were being produced in Birmingham, UK (Basalla, 1988). What is critically required for the Darwinian process is that these variants are of a kind that will *compete* with each other, and here too human culture exhibits the relevant characteristics. For instance, while there are 6,800 languages spoken worldwide and over 10,000 different types of religious belief, most individuals will predominantly speak only one language and follow one religion. Clearly, the struggle between cultural traits is played out not in terms of 'life' and 'death' but for limited 'slots' within human minds.

Unlike genes, the *inheritance* of cultural traits need not occur from parent to offspring, and stable transmission can follow a number of routes. Nonetheless, several studies have found that the attitudes of parents and offspring are often rather similar, and (even allowing for the confounding of genetic inheritance) the most obvious explanation for this outcome is that children learn attitudes socially. For instance, a study of Stanford University students revealed that the religious and political attitudes were strongly consistent between parents and offspring (Cavalli-Sforza et al., 1982). The same has been reported to apply in non-industrialized societies; among the Tsimane forager-farmers of Bolivia, the transmission of knowledge and skills generally occurs from older to younger generations (Schniter et al., 2023), and among horticulturists in the Democratic Republic of Congo, youngsters acquire knowledge about food taboos primarily from their parents (Aunger, 2000b). As more and more studies are conducted, broad cross-cultural patterns are starting to emerge concerning how children learn subsistence skills and norms (Lew-Levy et al., 2017, 2018), including through storytelling (Sugiyama, 2017), and Cultural Evolution models have indicated that learning from parents is particularly advantageous in stable environments (McElreath and Strimling, 2008). The importance of learning from the previous generation is reinforced by historical records and statistical analyses that report the consistent use of farming practices, modes of subsistence, social systems, and social preferences over hundreds, even thousands, of years (Durham, 1991; Mathew and Perreault, 2015). As noted later in the chapter, cultural transmission can also occur within generations.

These three basic characteristics (variation, competition, and inheritance) are necessary and sufficient to generate natural selection. If culture exhibits variation, if these variants compete with differential fitness, and if the variants are inherited, then cultural evolution, it is argued, *must* ensue through 'cultural selection'. Cavalli-Sforza and Feldman (1981) defined *cultural selection* as a process by which particular socially learned beliefs, or pieces of knowledge, increase or decrease in frequency due to being adopted by individuals at different rates. Of course, *natural selection* can also change the frequency of a cultural trait, through the differential survival of individuals expressing different types of traits, and cultural traits can also affect reproductive success; for instance, some religions proactively endorse large family sizes among their followers. Working with biological and cultural processes simultaneously has led researchers to account for how maladaptive cultural traditions might also evolve. When it has sufficiently high cultural fitness, cultural information could increase in frequency despite decreasing genetic fitness. For instance, in developed countries, fertility control (e.g. via contraception) is almost certainly at a disadvantage in terms of decreasing genetic fitness, based on the assumption that users would have fewer offspring, but contraceptive use has perhaps spread by virtue of its advantage in cultural selection.

Cultural Evolutionists suggest that the process of cultural selection explains why several other phenomena observed in biological evolution are also evident in human culture (Table 6.1), including *adaptation*, *extinction*, and *convergent evolution* (Mesoudi et al., 2004). *Cultural adaptation* is manifest in the complex, multifaceted, and ever more efficient solutions of technology and engineering. Basalla (1988) amassed extensive historical evidence for the

Table 6.1 Some parallels between biological and cultural evolution

Biological evolution	Cultural evolution
Discrete heritable units (genes)	Discrete heritable units (cultural traits, memes)
Genetic inheritance	Cultural transmission
Natural selection	Cultural selection
Genetic adaptation	Cultural adaptation
Extinction	Loss of a cultural lineage
Convergent evolution	Independent invention
Admixture and hybridization	Cultural mixing (e.g. creole languages)
Mutation	Innovation
Random genetic drift	Random cultural drift
Horizontal gene transfer	Borrowing across cultures
Genetic clines	Cultural clines (e.g. dialects)
Speciation	Split in cultural lineages
Fossils	Ancient texts/artefacts

Based on Pagel et al. (2007) and expanded.

gradual accumulation of technological modifications through time, such as Joseph Henry's 1831 electric motor, which borrowed many features from the steam engine, or Eli Whitney's 1793 cotton-sifting machine, which was based on a long line of Native American devices. Cultural traits go *extinct* as a result of competition or drift, as occurred with the gun in Japan in the seventeenth century and with bone tools in Tasmania around 3,500 years ago (Henrich, 2004; Perrin, 1980). Selection is also apparent in the *convergent evolution* of similar cultural forms in unrelated lineages, such as the tendency for both cartoon characters and teddy bears to become increasingly neotenous (baby-like) over the course of time (Gould, 1980; Hinde and Barden, 1985).

The parallels between the processes of biological evolution and cultural change lend support to the Cultural Evolution approach, whereby methods and theoretical frameworks from evolutionary biology are applied to the study of cultural entities (Mesoudi et al., 2006). Yet, while there are many fundamental similarities between biological and cultural change, the two processes are certainly not identical, and biological methods and models are rarely applied to cultural phenomena without consideration of potential differences (Boyd and Richerson, 1985). For instance, unlike genes and genetic mutations, cultural traits can sometimes blend together, cultural variation can be continuous (e.g. level of interest in current affairs) or discrete (e.g. smoker versus non-smoker), and cultural innovations are frequently non-random and guided by the goals of the innovator. However, it is important to recognize that Cultural Evolutionists do not unthinkingly apply genetic models to cultural data, but rather build bespoke models incorporating details of cultural transmission as core assumptions. We return to the potential differences between genetic and cultural evolution later in the chapter, and, in the following chapter, we evaluate the assertion that cultural evolution occurs at a much faster rate than does genetic evolution.

Types of cultural learning

Cultural Evolutionists have described a number of evolved learning rules, known as 'social learning strategies' or 'transmission biases', that specify when an individual should copy and from whom they should learn. These strategies include 'copy the most successful individuals', 'copy in proportion to the demonstrator's payoff', and 'conform to the majority' behaviour. Each generates a different pattern of cultural selection. Some of the best studied types of cultural learning are discussed below.

Given a choice been two alternative behaviour patterns, individuals may be more likely to adopt one variant than another (Cavalli-Sforza and Feldman, 1981). Boyd and Richerson (1985) refer to this outcome as *biased cultural transmission*. Various types of bias may exist. Through *direct bias*, individuals choose which of two or more alternative traits to adopt according to some intrinsic quality of the trait (also known as a 'content bias'). A direct bias might result from a genetic predisposition to favour certain types of information, perhaps manifest as differences in the memorability or effectiveness of the variants. This form of learning is similar to Lumsden and Wilson's 'epigenetic rules' or to how Evolutionary Psychologists think about learning. Genetically biased transmission is likely to generate adaptive behaviour much of the time (Richerson and Boyd, 2005). However, other forms of cultural learning are less consistent with the thinking of sociobiologists and Evolutionary Psychologists. For instance, the cultural traditions that an individual picks up will often depend upon who else in the population has adopted that tradition. In the case of *frequency-dependent bias*, the commonness or rarity of a behaviour affects the probability of information transmission. When, as often seems to be the case, individuals are disproportionately predisposed to adopt the behaviour of the majority, this positive frequency-dependent bias generates *conformity*. Conformist transmission has some potentially interesting consequences; for instance, it could result in a viable form of group selection, and it could lead to maladaptive outcomes. People may also use cues about one trait (e.g. sporting prowess) to choose which individuals to observe to acquire information about another trait (e.g. clothes fashions). This form of learning is called *indirect* (or 'model-based') *bias*. Conformist transmission and indirect bias are examples of 'context biases', where the adoption of cultural knowledge is contingent on the social context of learning (Henrich and McElreath, 2003).

The cultural traditions of a population may change over time if individuals alter the cultural information that they receive before passing it on. Boyd and Richerson (1985) discussed *guided variation*, which refers to a process by which individuals acquire information about a behaviour from others, and then modify the behaviour on the basis of their personal experience. Here, cultural variation is guided by individual experience, which may allow behavioural traditions to evolve gradually towards the optimal behaviour for that environment, as Human Behavioural Ecologists envisage. The transformations to cultural knowledge that individuals bring about can themselves stabilize traditions, particularly when individuals are biased by genetic predispositions, a phenomenon known as 'cultural attraction' (Scott-Phillips et al., 2018; Sperber, 1996), as discussed later in this chapter.

Cultural Evolution researchers have demonstrated that not all social transmission is biased. Recent theory reveals that even random copying is surprisingly adaptive under a broad range of circumstances (Rendell et al., 2010). This outcome is partly because the copied individuals have already undertaken some selection by choosing to perform the best actions from their repertoire, associated with the highest returns. Random copying can also arise in situations where the choice between different cultural variants has little impact on the chooser's biological fitness, generating random *cultural drift*. For instance, despite all of the care that parents put into selecting their child's name, collectively, parents behave in a manner that is identical to the case in which they had chosen names at random (Hahn and Bentley, 2003). The reason for this phenomenon is nothing more than the fact that common names are more likely to be observed and considered by parents than obscure names, and the likelihood that a name is chosen is roughly proportional to its frequency at the time. Similar patterns characterize a range of other cultural phenomena, including the decorative motifs on pottery, the popularity of dog breeds, and the citation of scientists in academic journals (Herzog et al., 2004; Shennan and Wilkinson, 2001; Simkin and Roychowdhury, 2003).

Cultural Evolution researchers are also interested in how information spreads within populations. The *mode* of transmission describes the route by which cultural knowledge passes among individuals (Cavalli-Sforza and Feldman, 1981), and different models are required for alternative modes of information transmission. As noted earlier, social transmission can occur *vertically* (i.e. from parents to offspring). However, learning can also occur *obliquely* from the older to the younger generation through non-parental routes (for instance, learning from teachers or religious elders) or *horizontally* within generations (such as learning from friends or siblings). Of course, genetic inheritance is primarily vertical, and, hence, as social transmission frequently occurs through some combination of these modes of information transmission, cultural evolution may commonly exhibit quite different properties from biological evolution. The range of modes of information transmission, and the diversity of alternative context-dependent social learning strategies, are thought to underpin the complexity of human culture (Rendell et al., 2011).

Case studies

In this section, we describe three case studies that illustrate the breadth of Cultural Evolution research. We first describe the use of laboratory experimental investigations of cultural evolution. We go on to show how the application of phylogenetic statistical methods has been used to investigate the evolution of language. Finally, we discuss cultural evolutionary explanations for the emergence of complex human societies.

Experimental investigations of social learning and cumulative culture

While early in its existence Cultural Evolution was an almost exclusively theoretical sub-field, a vigorous laboratory experimental tradition has developed over the past two decades. These include investigations designed to test some of the assumptions and predictions of the models. Some of the earliest studies in this tradition were carried out by Tatsuya Kameda and colleagues at Hokkaido University, Japan. For instance, in a simple computer-based task, Kameda and Daisuke Nakanishi (2002, 2003) explored the relative merits of human subjects' reliance on social versus asocial (individual) learning, a topic that had been subject to extensive theoretical investigation, and for which there were clear theoretical predictions. The computer-based task required participants to locate a rabbit in one of two locations, either through asocial learning, at a cost, or by copying the decisions of others, for free. As predicted by earlier theory, an equilibrium was reached in which individual learners and social learners coexisted, and the frequency of social learning increased when the cost of asocial learning was increased by the experimenters. In addition, groups in which individuals were allowed to exhibit both social and asocial learning outperformed groups that were restricted to using only individual learning. This experimental evidence was consistent with the prediction that social learning enhances average fitness in a group when individuals can strategically switch between social and asocial learning (Boyd and Richerson, 1995).

In a similar computer-based experiment, participants were required to choose which of two crops ('wheat' or 'potatoes') to plant on their hypothetical farms over repeated seasons (McElreath et al., 2005). The participants worked in small groups and were allowed to learn from their own experiences in the game, to copy a single other individual in the group, or to copy from the entire group. As predicted from formal theory, the amount of reliance on social learning increased with task difficulty, which in this case was the difficulty in discriminating between the yields of the two crops. However, other aspects of the results were unanticipated by theory. Of particular surprise, when given the opportunity, many individuals did not engage in any social learning, even under conditions where social learning would have increased their returns. In addition, *conformity* (copying the behaviour shown by the majority more than would be expected by chance) was only deployed when the environment fluctuated, despite the general superiority of this strategy versus copying based on the probability of encounter rates alone. Subsequent research using a similar design indicated that, when participants experienced a spatial, rather than a temporal, change in the online setting (representing a 'migration' event), they relied heavily on conformity, which led them to adopt locally adaptive behaviour (Deffner et al., 2020). In conclusion, participants in experimental settings do not always employ social learning in the manner predicted by initial models (Morin et al., 2021), but Cultural Evolution researchers can, and do, then adapt their models to take these observations into account (e.g. Denton et al., 2023; Witt et al., 2023).

A follow-up study using a similar task also reported restricted use of conformity, but here evidence was found for two personality types among the participants (Efferson et al., 2008b). A subset of the subjects behaved as 'conformists', and a second group, whom they called 'mavericks', relied almost exclusively on asocial information. The 'mavericks' appeared to be missing out on useful information by avoiding learning from others in this setting. An experiment by Thomas Morgan and colleagues (2012) at the University of St Andrews, UK, shed light on this issue. In their study, participants were first required to decide whether a pair of three-dimensional shapes were the same or different, and then given the opportunity to revise their decision after seeing the judgements of other participants ('demonstrators'). Individuals were disproportionately likely to adopt the majority view only when they were uncertain about their own decision and when the number of demonstrators was high. This study provided evidence for conformist bias, but also revealed how this strategy can be masked by other learning biases that are simultaneously operating. Cultural Evolution researchers have examined how individuals vary in the extent to which they rely on social learning depending upon a range of variables, including age, gender, learning ability, and past experiences (Mesoudi et al., 2016; Muthukrishna et al., 2016). This variability is fully consistent with the assumption that social learning is underpinned by complex trade-offs at an individual level and the emergence of 'wisdom of the crowds' at a group level (Kameda et al., 2022).

Alex Mesoudi at the University of Exeter, UK, and Michael O'Brien at the University of Missouri, USA, carried out an experimental investigation of the cultural transmission of prehistoric arrowheads. In a computer game, participants had to design and test 'virtual arrowheads' in 'virtual hunting environments' and were given the option to use information about the success of other peoples' arrowheads. These arrowheads could vary with respect to their overall shape, width, thickness, length, and colour, and the virtual environments were set up so that there were several different combinations of traits that conferred high hunting returns (in evolutionary terms, there were multiple adaptive peaks). The researchers found that, despite experiencing repeated hunts in which participants gradually refined their own designs, individuals would frequently abandon their own arrowhead and copy the designs produced by others (Mesoudi and O'Brien, 2008). Moreover, participants tended to copy the arrowhead design of the single most successful hunter, even with respect to non-functional features such as colour. The results were consistent with Boyd and Richerson's (1985) model of *indirect bias* and potentially explain actual patterns of prehistoric arrowhead variation observed in the archaeological record (Bettinger and Eerkens, 1999).

A distinctive feature of human cultural evolution is its cumulative quality, with technology and understanding exhibiting repeated refinements and improvements over time. Christine Caldwell and Alisa Millen at the University of Stirling, UK, investigated *cumulative cultural evolution* using a 'transmission chain' design, in which participants complete a task, demonstrating their performance to the next individual in the chain, who then does the same, such that knowledge and skills can accumulate with time. Two tasks were chosen: building a tall tower out of uncooked spaghetti and modelling clay, and producing a paper aeroplane that could fly as far as possible. For both tasks, information accumulated within groups, such that individuals later in the chain produced designs that were more successful than those of earlier individuals. Deploying different conditions that allowed for imitation (copying behaviour), emulation (copying outcomes), and teaching (guided instruction), Caldwell and Millen (2008, 2009) found that each of these mechanisms was effective in supporting cumulative cultural evolution; conversely, the effective transmission of more complex tasks, such as stone tool manufacture, may require teaching and language (Morgan et al., 2015). In another transmission chain experiment, allowing participants to transmit causal knowledge was not necessary for cumulative culture, and sometimes even hindered progress (Derex et al., 2019), because, for more complex technology, people appear to devise myths about causation that can undermine performance.

Simon Kirby and colleagues (2008) at the University of Edinburgh, UK, adopted a similar approach to studying cumulative cultural evolution in languages. In a computer-based task, human participants were presented with coloured shapes that moved in particular ways across the computer screen. These moving shapes were given random 'names' by the experimenters (e.g.

a black square that moved up and down might be called a 'ki-he-mi-wi'), and the participants were required to learn the names of each stimulus. They were then asked to provide the names of both seen and unseen stimuli, and their answers formed an 'artificial language' that was used to train the next individual in the chain. The multi-syllable names were initially assigned arbitrarily to the stimuli, and the stimulus–name combinations evolved along the experimental transmission chains to become easier to learn and more orderly in their structure. Participants who were later in the chain made fewer errors, most likely due to the emerging structure of the artificial language (e.g. the names of all shapes that travelled in a circular motion might end up including the syllable 'poi'), effectively recreating characteristics of actual human languages. Structured artificial languages specifically emerged in situations where each new 'generation' consisted of naïve participants (Kirby et al., 2015), as these conditions favoured language that was both simple and expressive, or where group size was large and participants interacted with numerous others (Raviv et al., 2019). These experimental studies have provided a useful vehicle for exploring social learning and cumulative culture under laboratory conditions (Caldwell et al., 2016).

Language evolution

In *The Origin of Species* (sixth edition), Darwin predicted that, if one could reconstruct the tree of human evolution, one would have the best classification of human languages. If this prediction is correct, we would expect that an evolutionary tree of a set of human populations, reconstructed entirely from genetic data, would be highly similar to a linguistic classification of the same populations. There is good reason for expecting some correspondence, as events responsible for genetic differentiation are very likely to generate linguistic diversification as well. For instance, the partial physical separation of a single population into two populations is likely to reduce contact between them, which will contribute to both linguistic and genetic divergence. It is well known that geographical distance between populations is a good predictor of decreasing exchange of individuals and also of genetic diversity, and the same is true for linguistic diversity (Cavalli-Sforza and Wang, 1986; Cavalli-Sforza et al., 1992).

However, languages may be transmitted horizontally to a far greater extent than are genes, and historical records provide several examples of language replacements that would break the correspondence between linguistic and genetic trees. For instance, in Europe, the replacement of Celtic languages by

Latin in the territories of the Roman Empire, and of Latin by Anglo-Saxon in England, are well documented. Cultural Evolutionists have asked whether linguistic data exhibit tree-like structures (irrespective of whether these lineages are congruent with genetic relationships). If tree-like relationships exist among languages, then phylogenetic methods that are used in evolutionary biology to understand the relatedness of species could legitimately be used to reconstruct the historical relations among languages and test hypotheses concerning past demographic and dispersal events. Conversely, if languages change primarily through mixing, borrowing, and horizontal transfer between lineages, then tree-like structures will not be apparent, and other methods of analysis would be more appropriate.

Russell Gray and Fiona Jordan, while based at the University of Auckland, New Zealand, conducted a pioneering study of language evolution, in the process illustrating how phylogenetic methods applied to language could be used to test hypotheses about human prehistory. They focused on the Austronesian languages spoken in Southeast Asia and the Pacific Islands. Previous researchers had documented sets of words, or lexical items, for each of seventy-seven languages spoken in this area of the world. Gray and Jordan (2000) applied established phylogenetic statistical techniques, regularly deployed within biology, to ask whether the languages fitted a tree-like structure and to identify the best tree. The tests established that the language dataset had a significant phylogenetic signal to it; that is, it had the tree-like property characteristic of biological evolution rather than the interwoven matrix that excessive mixing of language elements would generate. Gray and Jordan then used this tree to test two competing hypotheses for the colonization of the Pacific by Austronesian-speaking peoples. This colonization is known to have taken place in two phases, an initial expansion of hunter–gatherers from South-East Asia around 35,000 years Before Present (BP), and a second wave of Austronesian-speaking farmers from China and Taiwan around 5,500 years BP. Controversy has surrounded whether the Austronesian expansion was rapid, with little mixing with earlier settlers (the 'express-train' hypothesis) or slower and with more intermixing between the newcomers and the indigenous populations (the 'entangled-bank' model).

To test between these two models, Gray and Jordan first used archaeological data to predict the sequence of colonization from Taiwan through to New Zealand and Hawaii, showing that it had occurred in an ordered geographical pattern, then compared this pattern to the language tree. A good correspondence between the two would require few colonization steps to map the geographical evidence onto the language tree, while a poor fit would require many more steps. In other words, a rapid expansion of the population across the region would be expected to produce a distribution of languages that mirrored the geographical pattern, and hence the archaeological data, while a slower expansion that included mixing of languages with those of indigenous populations would produce a more complex picture, with a less clear fit between the archaeological data and the language tree. The obtained fit was very close, indicating that the 'express-train' model fit the language tree exceptionally well, and significantly better than would be expected by chance. In contrast, they found no support for the 'entangled-bank' model. Similar methods have since supported a number of hypotheses about human prehistory, including the hypothesis that the Indo-European languages expanded with the spread of agriculture around 8,000 years BP (Heggarty et al., 2023) and that the initial divergence within the Sino-Tibetan language family also occurred around 8,000 years BP, coinciding with the emergence of millet farming (Zhang et al., 2020).

Darwin's intuition that languages evolve in similar manner to biological evolution is supported by the evidence that language diversity can be successfully understood using phylogenetic methods. These methods can also shed light on the mechanisms of language evolution. For instance, Mark Pagel and colleagues (2007) at the University of Reading, UK, noted that some words, such as 'tail', evolve rapidly to have diverse forms in different languages, while others, such as 'two', evolve slowly, and have related forms across language groups. However, no compelling explanation had been given for this pattern. Deploying phylogenetic methods across eighty-seven Indo-European languages, Pagel and colleagues showed how the frequency with which words are used predicts their rate of replacement over thousands of years. Frequently used words evolved slowly, while infrequently used words evolved more rapidly, potentially as a result of random cultural drift (Newbury et al., 2017). The regular speaking of words appears to act like a purifying mechanism, preventing changes from occurring. Within English, words spoken at a higher frequency are more likely to be of Old English origin, and commonly used irregular verbs often retain their ancestral morphology. Some aspects of language have evolved slowly enough to allow homologous lexical forms to persist for tens of thousands of years, demonstrating that features of culture can have a replication accuracy as high as that of some genes. In addition, languages can evolve in punctuational bursts (Atkinson et al., 2008), mimicking patterns of macroevolutionary change reported in the fossil record.

Languages vary widely, but prominent linguists have argued that, around the world, linguistic change follows consistent patterns, reflecting constraints of universal linguistic structures in our brains (e.g. an innate 'language acquisition device'; Chomsky, 1965). Michael Dunn and colleagues (2011), at the

Max Planck Institute for Psycholinguistics in the Netherlands, used Bayesian phylogenetic methods applied to four language families to test this hypothesis, examining whether changes in word order followed a universal pattern. Contrary to their expectations, the team found that few word order features of languages were correlated, and that patterns of change were typically lineage specific. At least with respect to word order, it would seem that language families are evolving according to their own specific sets of rules, rather than following a universal pattern, and may represent alternative solutions to generating effective communication (Hahn and Xu, 2022). These findings support the view that linguistic structure is primarily determined by cultural evolution rather than being tightly constrained by genes. However, low-level cognitive biases might underpin some cross-linguistic aspects of grammar, such as the apparent preference for interpreting the first noun phrase in a sentence as the agent (Bickel et al., 2015) and the over-extension of meanings based on physical resemblance (e.g. a computer 'mouse'; Brochhagen et al., 2023). The current rapid loss of languages across the globe could, unfortunately, impede future research into language evolution (Skirgård et al., 2023).

The evolution of complex societies

Today, billions of humans cooperate and flourish in huge societies comprised largely of unrelated individuals, strikingly different from the small-scale societies that characterized much of our prehistory. The first large-scale, complex societies with extensive division of labour, significant differentials in wealth and power, and dedicated governance structures appeared approximately 5,000 years BP. How did these societies evolve from small groups in which cooperative endeavours could be stabilized by face-to-face interactions? Historians and social scientists have proposed numerous hypotheses for the emergence of large-scale societal structures, laying emphasis on the importance of population growth, trade, warfare, and several other factors. However, such analyses are generally formulated verbally and are focused on specific historical events. As a consequence, the causal mechanisms underlying these hypotheses, and the general patterns that might arise from them, are seldom uncovered. Cultural Evolution researchers have proposed that the emergence of large-scale cultural phenomena can occur through a process called 'cultural group selection'.

As discussed in Chapter 3, the sociobiological revolution was built upon a rejection of group selection arguments. Most evolutionary biologists now accept the findings of theoretical models suggesting that group selection is plausible though only under restricted conditions (Price, 1970; Uyenoyama and Feldman, 1980), yet many question how frequently such conditions arise. For example, one of the requirements for selection at the level of the group is that genetic differences between groups are maintained. However, the processes that uphold group differences and select between groups are typically weak compared with the processes that break down group differences and select within groups (Williams, 1966). For instance, genetic differences typically arise through 'drift' (i.e. random changes in the genetic composition of the group), but movement of individuals between groups will quickly erode these differences. Another problem is that any group that exhibits cooperation between individuals will be susceptible to individuals who cheat and gain the benefit without paying the costs, and these cheats are expected to thrive. For example, for altruism to arise as a group-level adaptation, groups of altruists would have to give rise to new altruistic groups significantly more frequently, or go extinct less frequently, than groups without altruists while somehow counteracting the influence of selfish interests.

The conditions for group selection are arguably more likely to be met when one cultural group exhibits socially learned behaviour or knowledge that places that group at an advantage over another group that exhibits an alternative behaviour or does not share that specific knowledge, known as *cultural group selection* (Richerson et al., 2016). Considerable evidence has been documented for group selection operating on cultural traits; it is not genes that are selected for but rather groups of individuals expressing a particular culturally learned idea or behaviour (Richerson et al., 2016). For instance, competition between religions, which commonly differ in their birth and death rates, could provide an example of group selection acting on cultural traits (Wilson, 2002). The claim has been made that, during the Roman Empire, Christians were more cooperative than non-Christians, assorted positively, and ostracized non-cooperators (Stark, 1997); if correct, this difference might have led to Christians historically having greater success in coping with resource scarcity and disease, resulting in higher survival, reproduction, and rates of conversion, and ultimately to the success of this religion. Similarly, it has been suggested that the expansion of conservative Protestant denominations in the United States has been mainly driven by variation in birth rates, rather than by rates of conversion (Hout et al., 2001).

Another branch of Cultural Evolution theory has made a contribution to understanding societal processes by incorporating cultural transmission into population dynamic and spatial ecology modelling frameworks. For instance, Peter Turchin at the University of Connecticut, USA, and his collaborators produced an ecological model that incorporated cultural evolution processes and used it to predict (retrospectively) where and when large-scale complex societies should arise (Turchin et al., 2013). The model's predictions were compared to historical data on the emergence of complex societies across Africa and Eurasia from 3,500 to 500 years BP, and the resulting model provided a good fit. In other words, the model was able to predict where and when complex societies would have emerged in our historical past. As anticipated by Cultural Evolution theory (Boyd and Richerson, 1985), complex societies arise where ultrasocial norms and institutions spread that increase the competitive ability of groups. Societies able to solve collective action problems, such as collaborating to thwart enemies or to provide systems of formal education, were able to grow and complexify.

As societies grow in size, the costly institutions that stabilize within-group cooperation are easier to maintain, thus providing an advantage to such groups. Those societies with traits that enable greater control and coordination of larger numbers would have typically outcompeted those lacking such qualities. Warfare is proposed to have been the primary force that led to the spread of ultrasocial traits (Turchin et al., 2013). Warfare intensity, in turn, was predicted by the spread through cultural transmission of military technology, such as chariots and cavalry, as well as by geographic factors such as the availability of wild horses for domestication. When the model was tested against the historical data, it gave a very good fit, explaining 65% of the variance in the data. Conversely, a model that failed to include the cultural transmission of military technology gave a poor fit to the data. Using a similar approach, Turchin (2003) was able successfully to sort between competing theories for the rise and fall of empires in Europe over the last 2,000 years, reaching similar conclusions, and, more recently, the approach has been applied to interpret contemporary societies, particularly the United States (Turchin, 2016). These investigations are provocative, as they support the suggestion, frequently disdained by historians, that there are general mechanisms that shape the course of history.

Turchin's analyses are also consistent with complementary work deploying phylogenetic tools. For instance, Thomas Currie at the University of Exeter, UK, and colleagues used phylogenetic methods to establish that the political complexity of South-East Asian and Pacific societies increased in a uni-linear sequence, from leaderless groups to simple chiefdoms, then complex chiefdoms, and then states (Currie et al., 2010a). Similarly, cross-cultural analyses have suggested that key aspects of human social organization, such as the presence of administrative hierarchies, record-keeping, and monetary systems, have coevolved in predictable ways across major regions of the world (Turchin et al., 2018). Although these findings are evocative of Herbert Spencer's discredited theory of societal evolution (discussed in Chapter 2), here societies are recognized as regularly transitioning back to less complex organizations. Cultural Evolution researchers have also addressed questions such as whether belief in moralizing gods might have stabilized interactions between groups (Lang et al., 2019) and whether punishment of non-cooperators might reduce the frequency of free riders (Currie et al., 2021). These researchers have also questioned whether contemporary ratcheting processes, such as greater environmental exploitation and increased global connectedness, could hamper collective attempts to curb over-exploitation and global climate change (Pisor et al., 2024; Waring et al., 2024). Understanding the cultural evolutionary processes that influence human societies and collective decision-making could be key to devising solutions to these issues (Ellis, 2024).

Critical evaluation

Cultural Evolutionists have been criticized by researchers who argue that our genes dictate which cultural information will be acquired and that cultural traits cannot be said to replicate, as the information undergoes reconstruction each time it is transmitted between individuals. Other researchers have questioned the parallels between biological and cultural evolution. Below we consider each of these arguments in turn. The critique that culture cannot be divided into discrete packages, which has been directed at both Cultural Evolutionists and Gene–Culture Coevolutionists, will be discussed in Chapter 7.

Do minds select culture?

One aspect of memetics, often viewed as sinister, was that human beings seemed to have been stripped of their ability to choose their own beliefs, values, and ways of life. Apparently, nefarious 'mind viruses' were running our lives. The memes were choosing and manipulating us, not the other way round. While beguiling, this surreal alternative perspective seemingly neglects the fact that our minds have evolved over millions of years. One might imagine that evolution would have fashioned us with an ability to evaluate alternative cultural traits and filter the available information that is adopted. If our bodies have an immune system to quell biological viruses, then should we not expect our minds to have analogous defences to suppress rogue memes? While Cultural Evolutionists do not all embrace the meme's-eye view, many equally have placed little emphasis on how our evolved psychology might shape our cultural learning. Might this stance be missing some of the underlying complexity to human behaviour?

According to Evolutionary Psychologists, minds are not like vacant apartments to rent, idly awaiting cultural knowledge to take up residence (Tooby and Cosmides, 1989a). The ideas, knowledge, and skills that we acquire from others are likely to reflect to a large extent our evolved predispositions. An individual's genes may bias them towards adopting one cultural trait rather than another, an example of 'content bias' (Boyd and Richerson, 1985). While some biases, such as the tendency to prefer sugar-rich foods or to find faces with large eyes attractive, may perhaps be universal, others, such as a liking for dairy products or alcohol, may vary across individuals according to their genotype. Some individuals are unable to drink milk in adulthood without becoming ill due to genetic variants that result in lactose intolerance. Similarly, individuals who carry particular variants of the genes that encode the alcohol dehydrogenase enzyme experience negative physical symptoms when drinking alcohol and are thus less likely to develop alcohol dependency. Behavioural geneticists report that genetic differences explain some of the variability in human behavioural and personality traits (Plomin, 2018), although these claims are thought to be inflated (as will be discussed in Chapter 7).

The claim that there are some heritable genetic differences among people that influence behaviour is nonetheless highly plausible. Most variable traits show substantial underlying genetic variation. Plant and animal breeders are usually successful when they conduct a programme of artificial selection to increase milk yield in cows or some economically important trait of cultivated plants, and this moulding would not be possible without genetic variation in the selected character. These results hold for behavioural traits in animals, such as the tendency of rodents to explore a cage or pigeons to return home. Given that the propensity of people to adopt a particular cultural trait could be affected by many aspects of brain chemistry and organization, and, given that it is likely that such aspects of the brain are affected by many different genes, it is credible that some of the variation in people's beliefs and behaviour is affected by genetic variation. Intuitively, it seems to make sense that individuals with a more intellectual bent will be more likely than others to take up chess, while sensation-seekers will be predisposed to go scuba diving or hang-gliding. If Evolutionary Psychologists are correct, many cultural traits will be found widely because selection has favoured particular 'content' biases. The empirical evidence supports the existence of content biases (Stubberfield,

2022), such as a bias towards transmitting information (e.g. stories) with emotional content (Eriksson et al., 2016; Heath et al., 2001).

It does not follow, however, that the field of Cultural Evolution equates to Evolutionary Psychology, as some commentators have suggested (Heyes, 2018; Lewens, 2015). Certainly, *some* Cultural Evolutionists appear close to Evolutionary Psychologists in suggesting that cultural transmission mechanisms 'represent a kind of special purpose adaptation constructed to selectively acquire information and behavior by observing other humans' (Henrich and Boyd, 1998, p. 217). Conversely, others represent the opposite pole of the spectrum in suggesting that 'distinctively human cognitive mechanisms are adaptive because they are shaped primarily by cultural evolution, not by genetic evolution' (Heyes, 2018, p. 219), and that the 'processes involved in imitation learning are domain-general' (Heyes, 2018, p. 132). In the middle can be found the sub-field investigating *social learning strategies* which, with respect to the origin of the capabilities underlying social learning, has been explicit from the outset in disavowing any commitment to 'specialized' or 'domain-specific' mechanisms (Kendal et al., 2018). Hence, there exists no field-wide commitment within the Cultural Evolution research community either to strong, 'innately specified' learning rules, or to weak 'innate structure'. The topic is very much a contemporary empirical issue.

Moreover, content biases are far from the only biases relevant to cultural evolution. Numerous examples of Cultural Evolution models show how cultural traits can spread through 'context' biases, even against the action of a genetically evolved predisposition or in the face of a fitness disadvantage (Henrich and McElreath, 2003; Kendal et al., 2018). Both empirical and theoretical findings suggest that, in some contexts, quite general evolved psychological mechanisms (such as 'conform', or 'copy the most successful individual') have been favoured, which seemingly do not greatly pre-specify information content (e.g. Morgan et al., 2012). Our capacity for cultural learning may be adaptive precisely because its biases are largely content-independent, because genes cannot predict which behaviour patterns will be advantageous in changing environments or novel social contexts. An understanding of why a cultural trait has been acquired may be better framed in terms of the question '*What is the local culture?*' or '*Who is the transmitter?*', than '*What is the genetic predisposition?*' Such social knowledge becomes essential in order to predict what information is acquired by the learner. Thus, while minds do select culture, this statement by itself does not tell us *how* the human mind creates complex social structures and cumulative culture. We return to this topic in Chapter 7.

Replication or attraction?

Other critics argue that there are no cultural replicators, since ideas are not transmitted intact from one brain to another (Morin, 2016; Sperber, 1996). Rather, from this perspective, the mental representations in one person's brain generate a behavioural output that is used by observers to construct their own representations in their own brains. Unlike biological evolution, where replicated genes are passed down the generations, these critics argue that cultural evolution doesn't involve the transmission of a 'replicator'. Rather the information is reconstructed in each new brain that it occupies. Here, we describe recent debates around whether cultural transmission involves replication or whether it is more helpful to consider that culture is reconstructed, then discuss how these debates might be resolved.

In terms of generating stable cultural transmission, the problem with reconstruction of information is that the observer's mental representation cannot be guaranteed to resemble the original, unless there exist biases, that predispose individuals to construct representations in similar ways. Dan Sperber (1996) called these biases 'cognitive attractors'. Where there is evolved structure in the brain that leads individuals to find a particular cultural variant attractive or salient, cultural 'replicators' might appear to exist when in fact cognitive or perceptual biases produced similar outcomes in each individual. Here, the stability of traditions is thought to come not from faithful replication but from consistent transformation (Scott-Phillips et al., 2018). The cultural transmission of eye gaze in portrait paintings serves as a useful example. Studies have reported that pictures of faces staring directly at the viewer are perceived by humans to be more attractive and attention grabbing than those with averted gaze (e.g. Ewing et al., 2010). Olivier Morin of the Central European University, Budapest, Hungary, reasoned that if a cognitive bias for direct-gaze images exists, then portrait paintings might have evolved over time under its influence to become increasing forward-facing. Sure enough, Morin (2013) found that Renaissance portraiture gradually changed to become dominated by direct eye gaze, with the same trend replicated in a Korean dataset.

There are at least three respects in which Sperber's viewpoint, referred to as 'cultural attraction theory' (Scott-Phillips et al., 2018), has been viewed as challenging for Cultural Evolution research. Sperber's argument was interpreted as problematic by researchers committed to the view that cultural evolution is critically dependent on replication, for instance, memetics (Dennett,

1995). However, to researchers open to the idea that many cultural traits are likely to have at least a degree of reconstruction or reliable transformation, this criticism does not obviously cause problems for the field. Indeed, Cultural Evolutionists seemingly have grounds to predict the existence of reconstruction. To the extent that cultural messages contain information that is likely to be accurately reconstructed by receiving brains anyway, then such messages contain redundant information. Natural selection eliminates redundancy, as organisms that do not waste energy and resources will generally outcompete those that do. Thus, only the cultural information that is absolutely necessary is expected to be transmitted, while the rest will be left for the receiving mind to reconstruct. Mesoudi and Whiten (2004) found exactly this pattern in a transmission chain experiment, where redundant information was systematically lost during transmission. Provided that reconstruction is socially contingent, that is, provided that the expression of a cultural trait by one individual provides the trigger for a second individual to produce the same trait, as advocates of cultural attraction theory maintain, then reconstructed traits are just as subject to cultural evolution as are transmitted traits.

A second inference that has been drawn from the above critique is that Cultural Evolution models, which commonly assume the existence of replication, discrete gene-like replicators, and cultural selection, will fail to describe cultural evolution accurately if the dynamics of cultural change are dominated by 'unconsidered' transformative causal mechanisms, guided by evolved domain-specific biases (Claidière and Sperber, 2007; Claidière et al., 2014; Scott-Phillips et al., 2018). However, whether Cultural Evolution theory does neglect transformative processes, attractors, or domain-specific biases has been contested (Henrich et al., 2008; Mesoudi, 2016). The process of *guided variation*, for instance, comprises the 'transformation' through individual learning of culturally acquired knowledge (Mesoudi, 2017). Few would contest that individual learning is shaped by evolved biases, that it can lead traditions to approach local optima (i.e. attractors), and that under such circumstances learning is a source of stability, and not just change. *Direct biases*, such as content biases, where individuals find one cultural variant more memorable, attractive, or effective than the alternatives, are another example of a well-studied cultural evolution mechanism in which reliable reproduction and stability of the cultural traits can be influenced by evolved cognitive biases. A key insight from the Cultural Evolution literature is that context biases can be a source of cultural stability, with processes like *conformist transmission* also generating population-specific traditions that serve as 'attractors', a point rarely emphasized by cultural attraction theory. We see little grounds for concern that contemporary Cultural Evolution theory fails to capture important mechanisms.

That said, this book lays stress on the value of pluralism in science. By encouraging researchers to consider the role of biased transformations in cultural evolution, cultural attraction theory serves as a useful counterpoint to the strong selectionist stance that has been present in the field of Cultural Evolution since its inception. Returning to the example of gaze in portraiture, the trend towards direct gaze can be accounted for by at least two processes. Morin (2013, p. 227) describes how the testimonies of young painters 'mention that they copied mostly the great masters of their day', a tendency that would generate a *success bias* (Henrich and McElreath, 2003), corresponding to a selective explanation. However, the trend is also compatible with the suggestion that 'young painters transformed their elders' style while copying it' (Morin, 2013, p. 227), consistent with a *guided variation* mechanism, which is a transformative explanation. Both types of explanation surely merit consideration. The cultural attraction challenge might even be interpreted as good news for Cultural Evolutionists, because, if humans are predisposed to reconstruct particular cultural traits, it will mean there is another reason to expect cultural stability to be high. The fact that cultural transmission might involve reconstruction, perhaps according to the directive of evolved genetic predispositions, does not detract from the hypothesis that culture evolves. The degree to which cultural evolution is reliant on selective or transformative processes is wide open for study (Acerbi and Mesoudi, 2015), and opportunities for rapprochement have been identified (Mesoudi, 2021).

Thirdly, cultural attraction theory has been viewed as a critique of the claim that cultural evolution is reliant on high-fidelity copying (Sperber, 1996; Sperber and Claidière, 2006). However, Cultural Evolution theory is not built on the assumption of accurate information transmission (Henrich et al., 2008). For illustration, Henrich and Boyd (2002) constructed a cultural transmission model with low fidelity copying of a continuous trait, and with weak cultural selection in the face of strong cognitive attractors. Despite these assumptions, the resulting population dynamics and the final distribution of the cultural trait were closely approximated by a discrete-trait replicator dynamics model. This finding implies that it may frequently be reasonable to assume the existence of cultural replicators in formal analyses, irrespective of whether cultural evolution is actually underpinned by replication, because it leads to accurate approximations (Henrich et al., 2008; Mesoudi, 2016). Henrich and Boyd (2002) further demonstrated that cultural replicators are not a necessary prerequisite for cultural evolution to occur. Any process of cultural transmission that leads to accurate reproduction of the characteristics of the population can generate cultural evolution. In sum, cultural change can be effectively studied using the mathematical techniques devised by Cultural Evolutionists irrespective of whether fidelity is high, or cultural replicators exist.

How similar are biological and cultural evolution?

Cultural Evolutionists argue that borrowing Darwinian concepts and methods, suitably adjusted to the structural peculiarities of human culture, is the quickest and easiest path to a reasonable theory of human culture and thus to an improved understanding of human behaviour (Boyd and Richerson, 1985). But this reasoning will only hold if the parallels between genetic and cultural processes are sufficiently robust. Here, we examine some of the apparent differences between cultural and biological evolution.

Various researchers have claimed that, unlike biological evolution, cultural evolution involves the merging of lineages (e.g. two music traditions might merge over time to create a new musical form) and the horizontal transmission of information within, as well as between, generations.

> Biological evolution is a system of constant divergence without subsequent joining of branches. Lineages, once distinct, are separate forever. In human history, transmission across lineages is, perhaps, the major source of cultural change. (Gould, 1991, p. 65)

In other words, whereas genetic lineages can be represented as bifurcating and tree-like, cultural lineages perhaps have more intertwined branches. Gould's point, that cultural information can cross conceptual lineages, was regarded by meme enthusiast Daniel Dennett (1995) as particularly troubling for memetics, for two reasons: first, because lines of descent become hopelessly muddled, and, second, because the outward expression of memes changes so fast that there is no chance of keeping track of particular memes over time. These concerns were sufficiently onerous for Dennett to express pessimism over whether a science of memetics could become established. Do the same concerns apply more generally to Cultural Evolution theory?

In fact, this is an instance where an overly simple model of biological evolution is employed as a standard against which to dismiss a 'complicated' notion of culture. To elevate this contrast to a dichotomy is a distortion of both biology and culture. That cultural evolution occurs predominantly through convergence of lineages is an assumption. Examining this idea empirically, a study of Turkmen textile artefacts found that fabric designs are better characterized as transmitted in a diverging tree-like manner rather than the converging pattern expected if there is extensive borrowing across lineages (Tehrani and Collard, 2002). Other examples of cultural traits that are stably transmitted down lineages have already been discussed in this chapter; language evolution, for instance, frequently exhibits the tree-like structure characteristic of vertical transmission. Conversely, convergence frequently occurs across biological lineages. For instance, lichens are compound plants consisting of both algae and fungi, while both mitochondria and chloroplasts evolved through cells taking in unrelated life forms. Genetic material can be transmitted across species boundaries through hybridization processes, through the actions of viruses and plasmids, and through the existence of transposable elements ('jumping genes') (Schaack et al., 2010; Taylor and Larson, 2019). There is increasing evidence that horizontal gene transfer is widespread among multicellular organisms (Schaack et al., 2010). These observations render the difference between cultural and biological evolution less distinct than implied by Gould's assertions.

That said, there may well be quantitative differences in the degree of merging between lineages observed among biological and cultural inheritance systems, and this cross-lineage transfer potentially creates problems for some (not all) Cultural Evolution theory. For instance, there are serious challenges associated with the application of phylogenetic methods devised on the assumption of vertical transmission and a tree-like structure to cultural phenomena, which researchers would be rash to ignore (Evans et al., 2021). In our judgement such concerns are best regarded as points for consideration, or spurs for the development or utilization of superior tools, rather than reasons to abandon current methodologies altogether. Theoretical studies have established that phylogenetic estimates are frequently robust in the face of realistic levels of borrowing across lineages (Currie et al., 2010b; Greenhill et al., 2009), and new phylogenetic methods specifically designed for borrowing are emerging (e.g. Neureiter et al., 2022). The immediate challenge for Cultural Evolutionists is to identify cases where the transfer of cultural knowledge across lineages seriously violates the assumptions of the analyses and to use alternative methods in such instances, provided that the alternatives are demonstrably superior.

It is also frequently asserted that cultural evolution is Lamarckian, and sometimes this is in itself regarded as sufficient to invalidate it (Hull, 1982). The term 'Lamarckian' is typically employed to depict instances where acquired characteristics are genetically inherited, and this process has long been discredited as playing any role in biological evolution. However, recent years have witnessed a broadening of conceptions of heredity; the germline transmission of epigenetic marks and the transfer from parents to offspring of a wide variety of additional resources (including symbionts, hormones, nutrients, and antibodies) are now recognized to contribute to parent–offspring similarity (Bonduriansky and Day, 2018; Jablonka and Lamb, 2014). Social learning fits comfortably within this broader view of inheritance. If Lamarckian inheritance is interpreted as meaning that acquired characters are incorporated as changes in gene sequences, then clearly, cultural evolution is not Lamarckian. However, the term 'Lamarckian' is often interpreted more broadly, as occurring where acquired characters are inherited through other mechanisms. On this interpretation, social learning can be viewed as Lamarckian because individuals frequently inherit information that others have previously acquired or modified (an instance of 'guided variation').

Whether or not cultural inheritance is appropriately described as Lamarckian also depends on how the culture 'genotype' and 'phenotype' analogues are defined (Laland and Brown, 2002). The important point, however, is that, since no-one thinks that social learning occurs through changes in DNA sequences, the label 'Lamarckian' cannot cast a slur over Cultural Evolution research. Individuals certainly modify the information they acquire and whether one regards that as resulting from a Lamarckian process, a cultural mutation, or some alternative action, heritable cultural variants will still be subject to differential transmission and adoption. We suggest that digressions over Lamarckian inheritance are unnecessary. A related, but more important, point is that cultural innovation is sometimes directed and intentional rather than random. However, the fact that mutations at the cultural level are sometimes not random but *smart variants*, informed by exploration, experimentation, and experience, is not a weakness of the field of Cultural Evolution, but a subject ripe for investigation (Fogarty et al., 2015). Cultural evolution occurs any time that heritable cultural variation is differentially propagated. It does not matter if the variation is generated at random (Henrich et al., 2008); indeed, Cultural Evolution models have been devised that assume non-random invention (Fogarty et al., 2015). In fact, the literature on human creativity indicates that much variation in culture is *not* directed in this sense as innovation or discovery is often the result of trial and error or serendipitous events. Cultural Evolution models have the flexibility to take into account both the similarity and differences between genetic and cultural evolutionary processes.

Conclusion

The field of Cultural Evolution has come a long way since its inception. The practice of Cultural Evolution has itself evolved into a serious rigorous science, with productive empirical and theoretical wings, and its influence has grown steadily (Creanza et al., 2017; Mesoudi, 2017). Controversies remain, regarding the degree to which biological and cultural evolution are similar, and the extent to which the differences between them are problematic. There are tensions with Evolutionary Psychology, whose advocates tend to lay more emphasis on evolved content biases, and perhaps less emphasis on context biases, than Cultural Evolution theory does. We will revisit some of these points of contention in the final chapter. First, however, we will consider an approach that allows both biological and cultural evolution to be explored simultaneously. That approach is called Gene–Culture Coevolution.

7
Gene–culture coevolution

The final sub-field that we describe is the only one that explicitly acknowledges that both genes and culture evolve and, additionally, that genes and culture interact, thus modifying each other's selective environments. The roots of 'gene–culture coevolutionary theory' can be traced back to the time of the human sociobiology debates. As described in Chapter 3, after considerable controversy around the emerging field of sociobiology and its application to human behaviour, Edward Wilson acknowledged that culture needed to be incorporated into any accounts of human social behaviour and organization. In collaboration with Charles Lumsden, Wilson developed mathematical models that explored the relationship between genes and culture, based on the assumption that whether an individual adopted a particular cultural element depended upon the characteristics of that individual's developing brain and therefore their genes (Lumsden and Wilson, 1981). Wilson was convinced that the study of gene–culture interactions would be a key area of future research, stating:

> It is possible that gene–culture coevolution will lie dormant as a subject for many more years, awaiting the slow accretion of knowledge persuasive enough to attract scholars. I remain in any case convinced that its true nature is the central problem of the social sciences, and moreover one of the great unexplored domains of science generally; and I do not doubt for an instant that its time will come. (1994, p. 353)

However, Wilson's contribution to this emerging field of research was relatively short-lived, as his book, *Genes, Mind, and Culture* (Lumsden and Wilson, 1981) was not well received by the academic community, partly because of his role in the sociobiology debates. In contrast, another book published in the same year by geneticists Luca Cavalli-Sforza and Marcus Feldman (1981), called *Cultural Transmission and Evolution*, gained a much more positive reception and is considered the foundational stone in this new sub-field of Gene–Culture Coevolution (henceforth, this phrase will be capitalized to denote this specific sub-field, whereas the phenomenon of 'gene–culture coevolution' will be denoted using lower-case letters).

In 1976, Feldman and Cavalli-Sforza had published the first simple dynamic models that incorporated both genetic and cultural inheritance. The innovative aspect of their work was that, in addition to modelling the differential transmission of genes from one generation to the next, they incorporated cultural information into the analysis, allowing the evolution of the two systems to be mutually dependent. One curious feature of the history of Gene–Culture Coevolution is that both archetypal sociobiologists (Wilson) and some of their most severe critics (Feldman and Cavalli-Sforza) almost simultaneously recognized the importance of gene–culture interactions and started to develop methods to address the problem. Feldman and Cavalli-Sforza were more cautious than Wilson in their application of these new ideas to human behaviour and had the benefit of substantial expertise in genetics and evolutionary modelling. Along with a raft of co-workers, Feldman and Cavalli-Sforza gradually built up an impressive body of mathematical theory exploring the interaction between genes and culture. As in their Cultural Evolution research, they frequently took advantage of the parallels between the spread of a gene and the diffusion of a cultural trait to borrow or adapt established models from population genetics. These researchers largely disavowed Lumsden and Wilson's work, and sometimes challenged their findings. As they had done for the sub-field of Cultural Evolution, Feldman and Cavalli-Sforza laid the initial theoretical foundations for the study of gene–culture coevolution, along with Peter Richerson and Robert Boyd (1985), who employed the term 'dual-inheritance theory', which we take here as synonymous with 'Gene–Culture Coevolution theory'. A new area of research had begun.

The questions addressed by Gene–Culture Coevolution researchers are of fundamental interest to the biological and social sciences: *'Do our genes restrict and delineate the nature of our culture?' 'What processes underlie human cooperation and conflict?' 'How did culture itself evolve, and how has it affected evolution in our lineage?' 'How do genes and culture influence the development of complex cognitive traits, such as intelligence?'* Gene–Culture Coevolution researchers treat culture as an evolving pool of ideas, beliefs, and knowledge that is learned and socially transmitted between individuals, while at the same time acknowledging that cultural learning is always reliant on biologically evolved knowledge-gaining structures in the brain. While initial research in this sub-field mostly involved mathematical modelling (albeit drawing heavily on empirical findings), dramatic advances in genetic techniques have given researchers an enormous amount of new data on the similarities and differences between the genomes of humans, the great apes, and our recently extinct relatives, such as Neanderthals. Gene–Culture Coevolutionists can now ask questions about behaviour, language, personality traits, and other topics (e.g. Kolodny and Edelman, 2018; Lotem et al., 2023; Uchiyama et al., 2021), based on the underlying assumption that genes and culture coevolve (Henrich, 2016; Laland, 2017).

Key concepts

Many of the key ideas associated with Cultural Evolution—a broad, pragmatic conception of culture, the exploitation of parallels between biological and cultural evolution, and the various forms of cultural selection—are also central to Gene–Culture Coevolution. Accordingly, in this section, we address concepts that are specifically relevant to Gene–Culture Coevolutionary theory. We will discuss the evidence that genetic evolution can be fast and, conversely, that cultural evolution can be slow. We will describe how humans have modified their selective environments through *niche construction*. Finally, we use a non-technical example to illustrate how Gene–Culture Coevolution models are built and discuss the insights that can come from generating such models.

Genetic evolution can be fast, and cultural evolution can be slow

Intuitively, we envisage culture as changing at a very rapid pace. For instance, relative to our grandparental or great-grandparental generations, we can point to dramatic advances in science and technology, such as the emergence of mobile phones, electric vehicles, and space flight, and advances in medicine, such as the proliferation of successful vaccines, diagnostics, and cancer treatments. Human communication can now take place over vast distances, and involve large audiences, through the invention of the internet and social media. The arts are continuously being reinvigorated, with new musical and artistic styles emerging that build on or replace preceding ones. Some changes, such as the rate of human-induced degradation of the natural world, are occurring at unprecedented speeds. The perception that changes in culture are inevitably fast, and that genetic evolution is slow by comparison, might suggest that these two strands of inheritance are working on different timescales and that cultural change therefore happens too quickly to influence genetic change. However, Gene–Cultural Coevolution researchers have instead argued that human evolution is dominated by a history of gene–culture interactions (Creanza and Feldman, 2016; Henrich, 2016; Laland, 2017; Richerson et al., 2010).

The assumption that genetic evolution always occurs slowly is no longer supported by the empirical evidence. As noted in Chapter 5, rates of genetic evolution in the animal kingdom are more rapid than was initially anticipated. For example, long-term field studies of birds and mammals, such as great tits, red deer, and savannah baboons, have revealed that substantial adaptive evolution can take place over relatively short time frames covering a few generations (Bonnet et al., 2022). The pace of evolution in some non-human animals is reported as being comparable with that documented for human cultural traits, such as changes in popular music and medical literatures (Lambert et al., 2020). Whether the high rates of genetic evolution observed in non-human animals are often matched in humans, and whether genetic changes, as opposed to plasticity, underlie phenotypic change observed in humans, remains to be widely established. Yet, these findings demonstrate that genetic evolution can take place at surprisingly fast rates, which opens up the possibility for genetic and cultural evolutionary processes to interact with each other on overlapping time frames. The assumed disparities in the rates of evolution of genes and culture perhaps reflect the fact that we often focus on short time frames when considering cultural change and long time frames when considering the evolution of lineages. Although cultural evolution might, on average, be faster than genetic evolution, microevolutionary rates of change could often be comparable for both genes and culture (Perreault, 2012).

In humans, cultural evolution is not restricted by generation time, given that cultural traits can be inherited from same-age peers or older individuals, as well as from parents, potentially allowing for the rapid spread of information and ideas. Yet, the archaeological record contains examples of cultural traditions that have remained remarkably stable over long time periods. For example, the earliest hominin stone tools (i.e. those of the Acheulian and Oldowan styles) retained their recognizable shapes and modes of production for hundreds of thousands of years. In the more recent archaeological record, microlithic tools that appeared in northern Asia around 18,000 years Before Present (BP) continued to be manufactured for another 4,000 years. In modern times, specialist craft activities often remain unchanged over the course of several centuries, and social institutions, such as labour markets, religious belief systems, the concept of democracy, and the use of currencies, can be traced back through long historical records. The general point here is that cultural evolution is not inevitably significantly faster than genetic evolution, although, as discussed later in the chapter, one of the adaptive benefits of culture might be its ability to track spatial and temporal variation in the environment. Gene–Culture Coevolution researchers can reasonably assume that the rates of genetic and cultural evolution are at least overlapping in their distributions to some extent. While culture may sometimes cause rates of environmental change that are too fast for human genetic evolution to track, there is ample evidence that genes and culture can coevolve (Feldman and Laland, 1996; Laland et al., 2010).

As in non-human animals, genetic evolution in human beings can be rapid. Large-scale genomic databases have provided indisputable evidence that human genomes have evolved in the recent past and are continuing to evolve to the present day (e.g. Voight et al., 2006; Wang et al., 2006). Sophisticated methods are now available for detecting signals of recent selection in human genomes, and such research has indicated that up to 10% of our genome has been impacted by selective sweeps over the past 50,000 years (i.e. late Pleistocene and Holocene) (Williamson et al., 2007). Some genetic variants (alleles) that have undergone recent selection exhibit variability between populations, including some involved in immune function, digestion, skin pigmentation, and response to specific climates, such as high altitudes (Fan et al., 2016). Exposure to pathogens, including as a result of cultural changes (such as the advent of Neolithic agriculture around 12,000 years ago), has left a particularly large impression on the human genome (Quintana-Murci, 2019), and exposure to new viruses (e.g. HIV and Covid-19) may continue to exert selection on genetic variants that provide even small levels of protection. Intriguingly, the interbreeding that occurred between modern *Homo sapiens* and their archaic relatives, such as Neanderthals, in some regions of the world might have provided selective benefits to modern humans in terms of increased pathogen resistance.

Gene–Culture Coevolution researchers propose that much of the recent selection on the human genome has resulted from culturally induced changes in the environment. For instance, selection on genes involved in immune response and digestion of specific food types is likely to have resulted from culturally initiated exposure to new pathogens, such as through close contact with domesticated animals, and changes in the diet (Quintana-Murci, 2019; Walker and Thomas, 2023). Genetic changes that are universally exhibited by all human beings (i.e. genes that have gone to 'fixation') can also be explained by gene–culture coevolution. For example, the *MYH16* gene, which is expressed primarily in the mandible and became inactivated in the hominin lineage, resulted in a reduction in the size of the jaw muscles (Stedman et al., 2004); selection on this gene coincided with the appearance of cooking in the archaeological record, which suggests that the learned ability to cook food selected for reduced jaw musculature. Similarly, increases in hominin brain size coincided with advances in technology (DeSilva et al., 2021). Genes expressed in the human brain and nervous system are well represented among recent selective sweeps. Later in this chapter, we evaluate the evidence that gene–culture coevolution has influenced human cognition and behaviour. Given that genome research had been biased towards studying individuals of European descent, a future challenge for genetics research is to determine which parts of the human genome have reached fixation and which exhibit variability.

Niche construction

If you were asked to describe the 'niche' of a particular organism, you might think about where it lives, what position it occupies in the food chain, or which aspects of the environment might explain its specialized features. We might think of species as competing with each other for particular niches. However, all organisms in part *construct* their own niches (Laland et al., 2000; Odling-Smee et al., 2003; Sultan, 2015). All forms of life, from unicellular organisms to fungi, plants, and animals, impact their environments, for example, through the extraction of resources and modification of nutrient cycles, through their choices of food and mates, and through the building of nests, webs, and burrows. The defining characteristic of niche construction is an organism-induced change in the selective environment; hence, the term also includes habitat selection, dispersal, and migration, whereby organisms relocate in space and experience new conditions. Through their niche-constructing activities, organisms influence the selection pressures acting on themselves and other organisms (Odling-Smee, 2024). Evolutionary biologists have investigated the effects of niche construction on the evolutionary process, showing that these activities can dramatically speed up, slow down, or change the direction of selection (e.g. Laland et al., 2017; Odling-Smee et al., 2003). Niche construction advocates have argued that ignoring the bidirectional interactions between organisms and their environments provides incomplete understanding of how evolution works.

Human beings are the most potent niche constructors on the planet. For instance, our recent cultural exploits have impacted all habitats across the world, with relatively little of the Earth's surface remaining untouched by our activities (Ellis et al., 2021). Over the last 50,000 years, modern humans have spread from Africa around the globe, experienced an ice age, witnessed rapid increases in densities, exploited agriculture, and domesticated hundreds of species of animals and plants. Each of these events represents a major transformation in the selection pressures experienced by humans and other species such as domesticates, and all (except the ice age) have been self-imposed. Looking further back, over the past 100,000 years, our hunter–gatherer ancestors lived in, and modified, a broad range of habitats and exhibited long-distance cooperation and resource management (Boivin et al., 2016; Singh and Glowacki, 2022). The field of archaeology has embraced a niche construction perspective, as this approach has provided new insights into human health, mortality risks, demography and life histories, as well as co-evolutionary responses in other species (Wells and Stock, 2020; Zeder, 2017). Such observations are consistent with the view that gene–culture coevolution has been a general feature of human evolution.

The significance of culturally induced niche construction (*cultural niche construction*) stems from the fact that it can be even more potent than niche-constructing capabilities that are based on genes (Laland et al., 2001). Some acts of niche construction, known as *inceptive* niche construction, initiate a change in an environmental factor and lead to changes in allele frequency. For example, by rearing domesticated animals, humans have experienced a new proximity to animal pathogens, creating selection for alleles expressed in immunity and resistance. Other cultural activities have dampened selection pressures; for example, building shelters and houses, and manufacturing clothes and heating systems, has lessened the selection on human morphology and physiology resulting from cold weather. In the absence of cultural niche construction, moving to a cold climate would lead to bouts of selection for genetic variants favoured in low temperatures. These activities, known as *counteractive* niche construction, oppose or nullify the effects of environmental change and protect organisms from shifts away from the environmental states to which they are adapted. Comparable forms of niche construction occur in animals, such as the habits of bees and wasps of cooling nests with water droplets and warming nests through muscular activity. Meta-analyses of selection gradients have confirmed that counteractive niche construction can reduce the strength and variability of natural selection (Clark et al., 2020).

The concept of 'niche construction' has been in the academic literature since the late 1980s. However, despite the evidence that considering niche construction has enhanced our understanding of human evolution, the importance of niche construction as an evolutionary process has not been universally accepted (Scott-Phillips et al., 2014). Some evolutionary biologists have suggested that the feedback between organisms and their environments is sufficiently incorporated into standard models of evolutionary theory and that the concept of niche construction does not provide any additional explanatory power (as discussed by Laland et al., 2014; Wray et al., 2014; see

Chapter 8). However, proponents of niche construction theory have pointed out that standard evolutionary models will sometimes produce the wrong answers if niche construction is not explicitly included as a causal process. The idea that niche construction is simply a product of prior selection acting on genes neglects the relevant empirical evidence showing that niche construction can influence the direction and strength of natural selection (Clark et al., 2020); this evidence demonstrates that causality is reciprocal. Moreover, such reductionism would be particularly problematic if applied to the study of human beings, whose environmental modification stems largely from cultural activities. Humans inherit genes, cultural knowledge, modified physical environments, and much more from previous generations, and the interactions between these inheritance systems are frequently bidirectional and coevolutionary.

Constructing Gene–Culture Coevolution models

The construction of Gene–Culture Coevolution models is a complicated procedure, and it is beyond the scope of this book to give a detailed account of this process. Instead, we will illustrate the logic behind the Gene–Culture Coevolutionary method with a deliberately simplified example (for a more technical introduction, see McElreath and Boyd, 2007).

Imagine ten friends, six men (Ahmed, Ben, Connor, Dan, Eric, and Filip) and four women (Aisha, Beth, Carla, and Dee). Several of these friends share a hobby of racing cars. *Racers* enjoy high-speed driving but are sometimes reckless and will take chances to win a race, and *non-racers* regard fast driving as frightening and prefer not to take part. Five of the friends are racers (Ahmed, Ben, Connor, Dan, and Aisha), while the others are not. In this example, racing (or not) is the cultural practice, and sex (male or female) is the genetic trait. Now what if Eric, under social pressure from his male friends, takes up racing? How could we track this change in racing behaviour within the population? For the moment, ignore any possible influence of an individual's sex on the propensity to be a racer, and let's look at what we would expect to happen by chance. We know that prior to Eric's switch the proportion of men among the friends was six out of ten (i.e. 0.6), and the proportion of racers was five out of ten (i.e. 0.5). From this, we would expect a specific proportion of the population, namely three of the ten individuals, to be racing men (calculated as $0.6 \times 0.5 = 0.3$). However, this was not the case, as four of the men were racers. By similar reasoning, after Eric has taken up racing, we would expect between three and four racing men (calculated as $0.6 \times 0.6 = 0.36$); however, there are in fact now five racing men. There is a discrepancy between the expected and observed numbers of each sex that are racers or non-racers. What is missing here?

The errors occur because of a non-random association between genes (in this simplified example, the sex chromosomes that we assume to determine the gender of an individual) and the cultural practice (racing or not). There is a departure from what we would expect by chance because a disproportionate number of men are racers. To describe the pattern among these friends accurately, we cannot just generalize from the overall proportions of each sex and each cultural practice in the group. Instead, we would have to track separately the gene–culture combinations, or the proportions of racing men, racing women, non-racing men, and non-racing women. The technical name for a particular combination of genotype and cultural practice is *phenogenotype* (Feldman and Cavalli-Sforza, 1976). The frequencies of each gene–culture combinations could be affected by two principal types of process. There are cultural selection processes, such as Eric's conversion, an example of horizontal cultural transmission that changed the proportion of racing men from four out of ten to five out of ten (0.4 to 0.5). There are also natural selection processes in operation. For instance, imagine that, while racing, Dan tragically dies in an accident, which drops the proportion of racing men back down to four out of nine (or, more precisely, 0.44). However, Filip and Beth become a couple and give birth to a baby girl who, initially at least, is a non-racer, which increases the proportion of non-racing females in the group from three to four (from 0.33 to 0.4). It would be relatively easy to construct mathematical expressions that describe the proportion of each gene–culture combination in the group each year, as a function of what they were in the previous year and how the frequencies have been changed by conversions to or from racing, as well as birth and death processes.

This example illustrates the logic of a Gene–Culture Coevolutionary analysis, although in practice such analyses are further complicated by the fact that, unlike for sex, the genetic component is heritable. While the changes in frequencies of genes and cultural knowledge can sometimes be modelled separately, in other cases, because of interactions, we need to keep track of the change in frequency of gene–culture combinations (i.e. phenogenotypes). Either way, in addition to the rules of Mendelian inheritance, transmission rules for cultural information (as outlined in Chapter 6) must be described. A common assumption is that the probability of an individual adopting a belief or preference depends on whether their parents have that belief, but equivalent models have been developed in which learning is either from unrelated individuals, from key individuals in the social group or from the majority in

Table 7.1 Transmission rules for the probability of children becoming racers or non-racers depending upon the racing behaviour of the parents

Behaviour of parents:		Probability of child being a:	
Mother	Father	Racer	Non-racer
Non-racer	Non-racer	b_0	$1 - b_0$
Non-racer	Racer	b_1	$1 - b_1$
Racer	Non-racer	b_2	$1 - b_2$
Racer	Racer	b_3	$1 - b_3$

the group. A set of such transmission rules is depicted in Table 7.1. Remaining with our racing example, we assume that, over the years, the friends pair up and have children. What are the probabilities that they will grow up to be racers or non-racers? If children are influenced by their parental teachings, then whether or not a particular child becomes a racer will depend, in part, on whether its parents were racers. We can set parameters that represent these probabilities of vertical cultural transmission, which are the b_i terms in Table 7.1.

Let us assume that children with two racing parents are more likely to develop an enthusiasm for racing than children with just one racing parent, who in turn are more likely to become racers than children with two non-racing parents. We can represent this in the analysis by simply setting the b_i parameters such that b_3 is greater than b_2 and b_1, both of which are greater than b_0. To give a specific example, this would occur if the children of two racing parents always become racers (so that $b_3 = 1$), children of two non-racing parents never become racers ($b_0 = 0$), and children with one racing parent become racers half of the time ($b_1 = b_2 = 0.5$). This case represents unbiased vertical cultural transmission and, if such rules apply, there would be no overall change in the frequency of racing behaviour as a consequence of cultural processes, although the frequency of racing might decrease as a consequence of natural selection when the occasional driving accident occurs. However, what if driving fast is so thrilling that racing parents can't stop talking about it? This situation might create a transmission bias, making it slightly more likely that children with one racing parent would become racers than non-racers (b_1, $b_2 > 0.5$). Now there are two conflicting processes that act on the frequency of racing: cultural selection favours racing and acts to increase its frequency, while natural selection favours individuals that don't race. Depending on the relative strengths of these two processes, racing behaviour may or may not increase in frequency.

Lastly, consider Daly and Wilson's (1983) hypothesis that the higher level of road accidents among men is a manifestation of a history of sexual selection in which males were selected for risk-taking strategies, while females were selected to be more risk averse. If that is the case, then the probability that a child becomes a racer depends not only on its parents' cultural practices, but also on its own genes (i.e. on its sex). Now the b_i parameters will take on different values depending on whether the child is male or female, and the frequency of racing will differ among the sexes, being higher in males. Even if a genetic predisposition toward racing is found among sons (b_{1m}, $b_{2m} > 0.5$) but not daughters ($b_{1f} = b_{2f} = 0.5$), the frequency of racing will reach higher than chance levels in females as well as in males. This is because there will be an increasing number of families in which at least one parent will be a racer, which will influence the chances of daughters as well as sons becoming racers.

Cultural transmission procedures such as those in Table 7.1, combined with rules for mating and genetic inheritance, allow researchers to derive a system of equations that describes how frequencies of genes and cultural practices change over time in the face of cultural selection, natural selection, and various kinds of interactions and biases. Equivalent sets of equations can also be produced for continuous traits, where the behaviour of an individual can be placed somewhere along a scale; for example, where individuals are described according to the average speed at which they drive. Gene–Culture Coevolutionary models allow researchers to ask questions such as, *'Could a predisposition for risk-prone behaviour be favoured in males?'*, *'Under what circumstances can cultural traits (such as driving fast) spread even if they reduce genetic fitness?'*, and *'What will be the final frequency of the (racing) behaviour in the population, when it reaches equilibrium?'* Mathematical models provide researchers with an understanding of processes that cannot be studied in other ways and can be a useful guide to empirical research, for instance, by generating testable predictions and highlighting key factors that researchers need to measure. In this manner, mathematical analyses can be indirectly evaluated using empirical data.

Case studies

In this section, we present three examples of Gene–Culture Coevolutionary research. We first describe evidence for the coevolution of the cultural practice of dairy farming and the genes that allow adult humans to digest milk. We go on to show how an explanation for human prosociality has been based on models of cultural group selection. We then describe a Gene–Culture

Coevolution perspective on the heritability of behavioural and cognitive traits, such as intelligence, which challenges the traditional views of behavioural geneticists.

Coevolution of dairy farming and lactose tolerance

The evolution of the ability of adult humans to consume dairy products represents a well-studied example of gene–culture coevolution. Unlike human infants, virtually all of whom can drink cows' milk without problems, adult humans vary considerably in their ability to digest milk. In fact, consuming large quantities of raw dairy products can actually make the majority of adult humans feel unwell. In such individuals, the activity level of the enzyme lactase in their bodies is insufficient to break down the sugar lactose in dairy products, and milk consumption can lead to sickness and diarrhoea. Whether or not adult humans can digest lactose is largely down to very small differences in regulatory genes upstream of the lactase gene, which cause the lactase enzyme to either continue functioning into adulthood, known as 'lactase persistence', or cease functioning after early childhood. A change in just one nucleotide, that is, one 'letter' of the genetic code, is responsible for lactose tolerance in Europeans, while several nearby nucleotide changes are associated with lactose tolerance in African and Middle Eastern populations (Ségurel and Bon, 2017). Animal milk provides an energy-rich food source that is associated with enhanced growth and health in children, potentially placing those individuals who are lactose tolerant at a nutritional advantage, particularly when other food sources are limited.

It turns out that a strong correlation exists between the incidence of the alleles for lactase persistence and a history of dairy farming in populations, with lactase persistence reaching a frequency of over 90% in some dairy farming populations but typically less than 20% in populations without dairy traditions (Gerbault et al., 2011). Farming and the domestication of animals for dairy products originated in the fertile crescent on the eastern edge of the Mediterranean around 10,000 years ago. Herding of domestic cattle, sheep, and goats then spread gradually through Europe, and northern and eastern Africa, over the following 4,000 years, along with the independent domestication of buffalo and zebu in southern and central Asia. Milk and processed milk products (such as yoghurt and cheese) have thus been a component of the diets of some human populations for over 6,000 years, roughly 250 generations. Is it conceivable that dairy farming created selection pressures that led to alleles for lactase persistence becoming common in pastoralist communities? Gene–Culture Coevolutionary theory provides a framework for answering this question.

Following work by Kenichi Aoki (1986), Feldman and Cavalli-Sforza (1989) used Gene–Culture Coevolution models to investigate the evolution of lactase persistence. They constructed a model in which the capacity to absorb lactose was affected by alleles of a single gene, with one particular allele allowing adults to digest milk without getting sick, and in which milk usage was a tradition learned from other members of the population. Their model showed that whether or not an allele allowing adult milk digestion achieved a high frequency depended critically on the probability that the children of dairy product users themselves became milk consumers (equivalent to the b_3 parameter in Table 7.1, if dairy product users are substituted for racers). If this probability was high, then a significant fitness advantage to the genetic capacity for lactase persistence resulted in the selection of the tolerance allele to high frequency. However, if a significant proportion of the offspring of milk users did not exploit dairy products, then unrealistically strong selection favouring tolerance was required for the allele to spread. In other words, differences in the fidelity of cultural transmission between cultures may account for genetic variability in lactase persistence. Crucially, there were a broad range of conditions under which the lactase persistence allele did not spread in the model despite a significant fitness advantage, indicating that traditional genetic models would have reached the wrong conclusion.

Other evidence also supports the hypothesis that dairy farming favoured the evolution of lactase persistence. Comparative statistical methods and computational analyses have been used to examine the incidence of dairy farming and alleles for lactase persistence across populations and have suggested that dairy farming evolved first, before the spread of the tolerance alleles, and not the other way around (Gerbault et al., 2011). Processed milk products, which have reduced levels of lactose, were probably consumed before raw milk became a staple dietary component in dairy farming populations, which suggests that the invention of milk fermentation, alongside with dairy farming, played a role in the gene–culture coevolutionary dynamics. The lactase gene shows one of the strongest signatures of recent selection in the human genome. However, some puzzles remain. For example, archaeologists have found that Neolithic sites in northern Europe, dated around 5,000 years ago, are rich in cattle remains and fragments of pottery with milk fat residues, and genetic data from cattle suggests that these livestock were being bred for increased milk yield during this period. Yet, analysis of ancient DNA from the fossilized bones of early Neolithic Europeans has shown that the frequency of the lactase persistence allele was low at this timepoint (Ségurel and Bon,

2017). The strength of selection for the persistence allele appears to have been greatest around 4,500–3,000 years BP in Europe and to have declined over the past fifty generations (approximately 1,500 years) (Mathieson and Terhorst, 2022). Additional factors, such as periods of famine and increased pathogen exposure, might have specifically advantaged those individuals who could digest raw milk products (Evershed et al., 2022). Although the details of how and when lactase persistence gained a selective advantage are still being revealed, the case for gene–culture coevolution of this trait remains strong.

This link between lactase persistence and dairy farming is only one of many examples in which culturally transmitted differences in the mode of subsistence and foraging behaviour, as well as agricultural and farming practices, have apparently influenced the selection of genetic variation among human populations (Laland et al., 2010; Richerson et al., 2010). Another example is the link across human populations between the consumption of starch-rich food and selection for the amylase enzyme, which is involved in starch digestion (Perry et al., 2007). Dogs also exhibit a similar evolutionary response in terms of selection acting on the amylase-coding gene, reflecting a shift in their diet to include more starch with their domestication in agricultural settings (Arendt et al., 2016). In humans, several genes related to the metabolism of carbohydrates, lipids, and phosphates also show signals of recent selection, including genes involved in metabolizing mannose, sucrose, and fatty acids, and the intracellular movement of cholesterol (Luca et al., 2010; Voight et al., 2006; Williamson et al., 2007). There is also evidence for diet-related selection on the thickness of human tooth enamel and bitter taste receptors (Kelley and Swanson, 2008; Soranzo et al., 2005). All such cases are probably manifestations of gene–culture coevolution and demonstrate the speed with which natural selection can leave a record in the human genome. Numerous genetic and comparative methods can detect signs of recent selection in the human genome, which suggests that additional examples of gene–culture coevolution are likely to be uncovered over time.

Human cooperation and prosociality

The previous case study focused on the role of gene–culture coevolution in generating variability across human populations. Yet, Gene–Culture Coevolutionary models also examine how some universal traits, along with the genes underpinning them, might have evolved. Comparisons between the genomes of our own species and those of our closest primate relatives have identified numerous lineage-specific changes in gene coding and regulatory mechanisms that have occurred since divergence from a common ancestor (O'Bleness et al., 2012), although linking these differences to human-specific phenotypic traits is difficult. In terms of behaviour, several lines of evidence indicate that human beings differ from other primates in their levels of cooperation and *prosociality*, defined as behaviour that benefits others (Henrich and Muthukrishna, 2021; Tomasello, 2023). This cooperation has resulted in the large-scale organizations and nation states of the post-industrial world, as well as the exchange relations, food-sharing, and cooperative hunting and childcare of hunter–gatherer populations. Such activities appear to rely on the fact that people place value on fairness, equity, and reciprocity, and are willing to punish those who do not act in a cooperative manner. Gene–Culture Coevolutionists have attempted to provide an answer to the puzzle of human prosociality based on the concept of *cultural group selection* that was introduced in Chapter 6.

Cultural group selection is the idea that cultural traits can evolve because of the advantage that they provide to the group; it may occur when one cultural group directly out-competes or is more successful than another because of the socially learned behaviours that the group exhibits. Cultural group selection has been proposed to provide a strong explanation for the evolution of human cooperation (Henrich and Muthukrishna, 2021; Richerson et al., 2016). Standard explanations of cooperative behaviour that rely on reciprocity and kin selection are insufficient to explain the extensive cooperative behaviour that is seen in both large-scale and small-scale societies. Gene–Culture Coevolution researchers instead make the case that a history of cultural group selection has favoured prosocial traits that stabilize within-group cooperation, such as a willingness to reward others for cooperative behaviour ('strong reciprocity') and punish those who transgress social norms ('altruistic punishment') (Fehr and Fischbacher, 2003). Cultural groups that exhibited these traits are argued to have outcompeted cultural groups that did not exhibit these behaviours, resulting in selection for genes underlying prosocial tendencies, and eventually leading to a universal trait of prosociality being exhibited across the whole species. Underpinning these behaviours is a willingness to follow society's social norms, rules, and values, often by conforming to the majority behaviour. This set of behaviours has been referred to as 'tribal social instincts' (Richerson et al., 2016), although not all Gene–Culture Coevolution researchers use this phrase, given the problems with the concept of 'instinct' (see Chapter 8).

One line of evidence that Gene–Culture Coevolution researchers have used to support this argument is the output of mathematical models and experimental studies. For example, Henrich and Boyd (1998) constructed models to investigate the evolution of conformity and found that virtually all of the circumstances that favour a reliance on social learning should also lead to conformity. Experimental studies have confirmed that, when individuals learn from others, they frequently tend to do what the majority of the population are doing (Efferson et al., 2008b; Morgan et al., 2012). Conformity makes it hard for new behaviour to spread within a population, as only common variants are favoured by cultural selection, which means that, if groups of individuals differ in their behaviour, conformity will act to maintain these differences while minimizing differences within groups. Theoretical work has also established that altruistic punishment can evolve, particularly when punishment is coordinated, for instance, through collective decision-making (Boyd et al., 2003, 2010). Thus, a combination of conformity and punishment of non-conformers could maintain sufficient group differences for cultural group selection to occur. In addition, symbolic group marker systems, such as human languages, cultural icons, totems, and flags, reinforce such differences. Arbitrary symbolic markers, though initially meaningless, become associated with predictable choices in multi-player cooperative games, leading to cultural group formation and ingroup favouritism (Efferson et al., 2008a). Such findings, which support earlier mathematical theory, increase the plausibility of cultural group selection as an explanation for human prosociality.

A second line of evidence for the link between prosociality and cultural group selection involves studies of between-group differences in cooperative and competitive behaviour. Such studies, often carried out by teams of anthropologists and economists in small-scale societies, have documented differences in how people respond in economic games. For instance, Joseph Henrich, now at Harvard University, USA, and colleagues (2001, 2004) recruited subjects from fifteen small-scale societies in twelve countries around the world and paid them to play economics games, such as the 'ultimatum game'. Considerable variation was found across societies in the level of prosociality, as reflected in the mean offers made and rates of rejection of those offers, suggesting that expectations are affected by group-specific conditions, such as local fairness norms. In a study in Kenya, cooperation between cultural groups was predicted by the level of cultural similarity (Handley and Mathew, 2020), and data from Western societies suggest that, in workplaces, levels of trust within groups are higher when between-firm competition is greater (Francois et al., 2018). Ancestral selection between cultural groups may also explain hostility and aggression to members of other groups, fear of strangers, slanderous propaganda concerning outsiders, and the evolution of warfare between groups comprising hundreds to millions of individuals (Bowles, 2009; Richerson et al., 2016; Zefferman and Mathew, 2015). It remains an intriguing possibility that some universal aspects of human social behaviour are the product of historical cultural group selection.

A related idea is that humans have undergone a 'self-domestication' process, whereby selection for enhanced prosociality has resulted in a cascade of other traits undergoing correlated selection (Hare, 2017; Sánchez-Viallagra and van Schaik, 2019). Domestication of farmed species (such as cows, sheep, and goats) has typically led to the emergence of a predictable suite of traits, such as increased tameness, smaller brain size, and shorter muzzles, and such traits also appear to be correlated in other species, such as urban foxes. This domestication effect is thought to involve changes in neural crest development in early foetal life, although contention remains over the centrality of this specific mechanism (Johnson et al., 2021). The challenge for applying a 'self-domestication' hypothesis to human evolution is that it is not clear which specific trait might have been favoured by selection and which traits followed along due to biases and constraints within the developmental process. During human evolution, selection might have favoured greater social tolerance, lower aggression, changes in craniofacial features, or even reduced tooth size, resulting in a whole suite of correlated traits changing over evolutionary time. Currently, there is little evidence that prosociality was the driver of that evolutionary response, as opposed to an indirect outcome. Whether the concept of 'self-domestication' provides a compelling explanation for the evolution of human prosociality and associated traits remains open to question.

Heritability of cognitive and behavioural traits

To what extent do genes and environments account for between-individual differences in cognition and behaviour? Is there a way to determine the relative importance of 'nature' and 'nurture' for complex traits such as human intelligence? The field of *behavioural genetics* attempts to answer these types of question (Plomin, 2018). Behavioural geneticists use a range of modern techniques to probe human genomes and search for associations between variations in the genetic sequence and occurrence of the specific trait or disease of interest (Young et al., 2019). Information from family lineages is used to seek out potential chromosomal regions or candidate genes that are related to particular disorders or diseases. Twin studies have provided an additional source of information; in particular, behavioural geneticists have used comparisons between monozygotic (identical) and dizygotic (non-identical) twins, on the assumption that both types of twins share the same rearing environment, whereas only monozygotic twins also share the same set of genes. For traits where monozygotic twins are more similar to each other than are dizygotic twins, the assumption is that genes play an important role. However, twin studies rely on the perhaps unrealistic expectation that both types of twins have equally similar experiences when growing up. The fact that dizygotic twins can be opposite sex and might look very different, and might be treated more differently from each other, makes this assumption unlikely to hold (Feldman and Ramachandran, 2018). Gene–Culture Coevolution researchers have more generally questioned how the 'environment' is incorporated into standard behavioural genetics frameworks.

Before progressing, though, it is important to understand the concept of *heritability*, which is broadly defined as the proportion of variance in a phenotypic trait across a population that can be attributed to genetic differences between individuals. Heritability scores can range from zero to 1, where a score of 1 is taken to indicate that all the between-individual variation in a trait is related to differences among genes. Yet, because learning from parents is often confounded with genetic inheritance, a high heritability estimate could actually reflect strong cultural transmission between generations; for instance, children born into a particular religious sect might share a set of defined behavioural traits with parents but not as a result of shared genes. Conversely, a heritability of zero arises when all individuals in a population share the same specific genotype and all differences in the trait of interest are therefore due to the different environments they experienced; for instance, the vast majority of people have ten fingers, with a heritability score close to zero, as most of the variation will reflect accidental loss of digits. These examples show that interpreting heritability scores is not straightforward and depends upon the choice of study population.

A low heritability score does not necessarily mean that genes are not involved in the expression of trait X, and, similarly, a high heritability score does not mean that trait Y is mostly dependent on genes and not much on the environment; such interpretations are wrong. Heritability is instead a measure of the proportion of between-individual variation in the trait that relates to genetic relatedness between those individuals in that specific population at that specific timepoint. For instance, heritability scores for reading in young children starting school were found to vary between Scandinavia, the United States, and Australia, with the lowest scores in Scandinavia (0.33 versus 0.68 in the United States and 0.84 in Australia) (Samuelsson et al., 2008); one year later, the heritability scores were similarly high in all groups. As Scandinavian nursery schools initially focus on social and emotional development, rather than reading, the variable amount of reading instruction at home was likely to have been more influential than genetic differences in this sample of preschool children.

The impact of decisions made by parents in home environments, and by society though the creation of institutions and social norms, confirms that 'culture' is an important part of the human 'environment'. Behavioural geneticists do acknowledge that the effects of specific genotypes might depend upon the environment that individuals with those genotypes encounter, and they incorporate a 'gene–environment' interaction term into their models to account for a proportion of the variation. This interaction term sits alongside terms in their models that specifically relate to the proportion of variability in a trait that relates to 'genes' and the proportion of the variability that relates to the 'environment'. The variation in the distribution of the relevant trait is thus divided into these three components. However, a Gene–Culture Coevolution perspective proposes that the *interaction* between genes and environment is the singular, overarching variable that should be investigated and that considering genes and environments as separate causal pathways relies on false assumptions (Uchiyama et al., 2022). In human beings, those aspects of the environment that relate to our culture are perhaps the most impactful, given the extensive cultural niche construction that our species undertakes. Gene–Culture Coevolutionists make the case that studying the interaction between genes and culture will provide a more complex, but realistic, view of heritability and will provide novel insights into previously intractable questions in behavioural genetics, as well as providing new predictions for how heritability scores are likely to vary between societies and across time periods (Muthukrishna, 2023; Uchiyama et al., 2022).

The study of the heritability of intelligence, often measured via an *intelligence quotient* (IQ), provides an example of the complexity of calculating heritability for a complex cognitive trait. Debates about the heritability of IQ can be traced at least back to the 1970s, when the 'intelligence debates' were raging both within academia and the general discourse. The educational psychologist Arthur Jensen (1969) suggested that racial disparities in average scores on IQ tests in the United States reflected heritable differences in abilities that educational and economic opportunities would fail to diminish. Richard Lewontin, one of the key architects of niche construction theory, immediately pointed out that heritability scores cannot be compared across groups without understanding the role of the environment, a point that was later reinforced by Feldman and Lewontin (1975). Numerous studies have revealed that country-level scores on IQ tests have risen over time in step with the advancement and adoption of formal education (Pietschnig and Voracek, 2015), which confirms that IQ is highly dependent on the cultural environment. High heritability score for IQ could indicate that educational factors are becoming more similar across individuals and groups, which would have the effect of 'unmasking' any remaining, potentially small, effects of genetic differences (Uchiyama et al., 2022). Unless educational, social, and physical environments are completely equal across individuals, heritability scores for IQ will always reflect the interactions between genes and environments, including those aspects of the environment that are culturally determined.

Examining the interactions between the inheritance of genes and the inheritance of culturally constructed environments will require more complex and sophisticated models than those traditionally used by behavioural geneticists. These researchers are aware of the potential challenges, and opportunities, that an interactionist perspective provides for their field (e.g. Turkheimer, 2000). If such a perspective were adopted, correctly attributing sufficient importance to cultural inheritance is likely to sharply reduce estimates of genetic contributions to phenotypes. For instance, heritability estimates for IQ of around 0.3 have been obtained using Gene–Culture Coevolutionary models applied to familial datasets (Otto et al., 1995), and these estimates contrast starkly with those generated using twin data alone, which typically range from 0.6 to 0.8. Such observations help to explain the general failure of behavioural genetics to find as much heritable genetic variation as would be expected on the basis of twin studies alone. In human IQ studies, for instance, single-nucleotide polymorphisms can, at most, explain 5% of variance in IQ performance (Sniekers et al., 2017). Evolutionary behavioural scientists have also pointed out the more general issues with measuring and comparing 'intelligence' between individuals and groups (Sear, 2022; Zefferman, 2023), highlighting the importance of accurate interpretations of heritability estimates.

Whether behavioural genetics will undertake the necessary changes to its methodology and epistemological framework remains to be seen, given the dominance of standard approaches (e.g. Plomin, 2018). The cultural environment of the genome is not a passive backdrop against which the effects of genes can be quantified, but a rapidly evolving frame of reference. Human beings are characterized by extreme amounts of cultural variation, which exceeds genetic variation by an order of magnitude or more (Bell et al., 2009). It is a real possibility that heritability scores have been seriously overestimated using standard methods, and that more sophisticated models for the inheritance of complex human traits that recognize the impact of Cultural Evolution are required in order to provide an accurate understanding of our genetic and cultural legacy.

Critical evaluation

Gene–Culture Coevolution research has been subject to many of the same criticisms as Cultural Evolution theory. Rather than repeating the points covered in Chapter 6, we will concentrate here on additional concerns that have been raised. First, we consider the argument that culture cannot be broken down and modelled as discrete units. Second, we evaluate the argument that the human brain is unlikely to be the product of gene–culture coevolution, because it is too complex an organ to have been subject to rapid selection. Finally, we consider the concern that the role of gene–culture coevolution has been overstated, either because culture is simply explained as a product of our genes or, conversely, because our advanced cognitive abilities are better understood as largely a product of cultural evolution.

Culture does not come in clean, discrete packages

For many social scientists, a major stumbling block for both Cultural Evolution and Gene–Culture Coevolution is the question of whether culture really can be divided up into units and modelled mathematically. Critics have suggested that culture cannot be separated into discrete packages with clearly specified boundaries, so its comprehension is not well served by the application of genetic approaches or models that assume discrete, well-bounded cultural units (e.g. Bloch, 2000; Ingold, 2007). Can the culture of a people be treated like a gene pool, that is, as a statistical description of the cultural variants present in a population? Is culture well characterized as a collection of 'beans in a bag', like population genetics refer to genes? Certainly, the current fashion within the social sciences is to lay emphasis on a more qualitative, holistic description of cultural phenomena.

Two arguments can be made in defence of the shared Cultural Evolution and Gene–Culture Coevolution perspectives. First, genes, too, are not as clean and simple as commonly conceived and, as discussed in Chapter 5 (where the similarities and differences between cultural and genetic evolution were evaluated), genetic evolution is not as straightforward as it might seem, yet treating genes as discrete units has proved extremely fruitful. If genes are regarded as clean, particulate pairs of alleles that reside at an easily definable locus on a well-charted chromosome and species are conceived as self-apparent natural kinds, then, in contrast, the boundaries to cultural traits will appear disturbingly fuzzy. However, cultural units only appear hazy in comparison with these misleadingly simplistic concepts of gene and species. The modern concept of the gene is characterized as abstract, general, and open, with fuzzy boundaries that change depending on the context in which the gene is expressed (e.g. varying across cell or tissue types). Multiple, mutually incompatible concepts of the gene can be found within biology (Rheinberger and Müller-Wille, 2018; Stotz and Griffiths, 2004). So, although the critics of Cultural Evolution are probably correct in pointing out the vaguely and flexibly specified nature of culture, the same problem applies to the gene concept, which undoubtedly has been of enormous value in the study of biological evolution. If uncertainty about the boundaries of a gene or species has not prevented progress in evolutionary biology, why should the fuzzy units of cultural evolution be regarded as problematic?

Second, viewing culture as a set of discrete packages is a pragmatic stance that potentially enhances our understanding of cultural phenomena. In both Cultural Evolution and Gene–Culture Coevolution studies, individuals are typically classified according to whether they possess a particular package of psychological constructs; for example, whether they believe dairy products are good to eat, know sign language, or prefer sons to daughters. This approach is conceptually no different from focusing on a particular gene and classifying individuals according to genotype. It does not mean that all other aspects of an individual's culture are irrelevant, but rather that it is instructive to consider the average effect of the particular information across the entire population. For Cultural Evolutionists, breaking down culture into units (or characterizing traits as a statistical distribution) is a useful theoretical stance. As described above, it is perfectly possible to model culture in other ways, for instance as a continuously varying trait (Cavalli-Sforza and Feldman, 1981; Henrich et al., 2008). Likewise, there is no reason why the transmission of multiple cultural traits cannot be modelled together as a 'package', as shown by those studying cultural evolutionary systems (Buskell et al., 2019; Yeh et al., 2019).

The bottom line is that biologists and human scientists alike will not be able to understand cultural processes unless they are prepared to break them down into conceptually and analytically manageable ways. Gene–Culture Coevolution researchers are fully aware that this stance downplays some complexities of human culture, but regard it to be a useful idealization for the scientific study of complex phenomena. Potentially, as empirical knowledge accumulates, the complexity of such theoretical models may also increase and can incorporate further nuances of cultural transmission, including more direct investigation of how multiple distinct cultural traits interact. At the same time, there is considerable and compelling evidence that humans acquire packages of learned and socially transmitted information, store them as discrete units, chunk and aggregate them into higher-order knowledge structures, encode them as memory traces in interwoven complexes of neural tissue, and express them in behaviour. It is perhaps not such an extraordinary claim that culture is acquired and expressed in packages.

Is the brain subject to gene–culture coevolution?

The idea that genes involved in the functioning of the immune system and digestive system have been influenced by culture is relatively easy to accept. We could imagine how culturally induced changes in farming practices or population density exposed our ancestors to novel diseases, with selection having favoured any genetic variants that provided protection, and likewise that changes in diets may have left signatures in the human genome. More challenging perhaps is the notion that genes involved in the functioning of the central nervous system have been impacted by gene–culture coevolutionary processes. Does this idea imply that between-group differences in cognitive and behavioural traits could be determined by genes? Does it bolster racist and deterministic accounts of the link between genes and human behaviour? Since the inception of this sub-field, proponents of Gene–Culture Coevolution have been vocal critics of racist interpretations of human genetic datasets (Feldman and Lewontin, 2008; Feldman et al., 2003), as such interpretations would be a serious distortion of how genes and culture coevolve. Yet, at the same time, Evolutionary Psychologists have focused predominantly on the shared, 'universal' aspects of human cognition (e.g. Tooby and Cosmides, 1990b, 2005), perhaps to avoid the challenge of addressing cross-population genetic variation. So how *did* our brains evolve, and what role did culture play?

The ancestral lineage leading to our own species, *Homo sapiens*, diverged from that leading to our closest living relatives—chimpanzees and bonobos—around 6 million years ago. Compared to the brains of these great apes, human brains are large relative to body size and are characterized by high levels of connectivity and developmental plasticity, which contribute to our advanced cognition and behavioural flexibility. Comparative genomic studies have started to document the differences in amino acid sequences between human and chimpanzee genomes, and such studies have revealed that hundreds of genes are differentially expressed in the brains of humans and those of chimpanzees (Pollen et al., 2023). These genes are involved in a range of neurodevelopmental processes, such as neurogenesis (i.e. the creation of new neural cells) and the growth of dendrites and synapses, which link brain cells and create new circuits (Zhou et al., 2024). Up to 5% of the human genome differs from the chimpanzee genome, equating to millions of genetic differences (Suntsova and Buzdin, 2020), although many of these differences are unlikely to contribute to phenotypic differences and are thought to be evolutionarily neutral. These lines of evidence suggest that human-specific variations in the genome, particularly in the organization of gene regulatory networks, have generated important changes in the size and connectivity of the human brain.

Additional insights into human brain evolution have been made possible by the extraction of genetic material from the fossil remains of our closest archaic relatives. Human beings are not the only species within the *Homo* genus to have inhabited the world. Our most well-known relatives, the Neanderthals (*Homo neanderthalensis*), split from the lineage leading to modern *Homo sapiens* around half a million years ago and thrived in Eurasia for at least 350,000 years. They became extinct 40,000 years BP, which is around 3,000–5,000 years after anatomically modern *Homo sapiens* arrived on the continent from Africa. This temporal and geographical overlap is now known to have resulted in interbreeding between Neanderthals and our own species (Gokcumen, 2020). In 2008, fossil evidence for a sister species of the Neanderthals, named the Denisovans, was discovered at a site in southern Siberia, and genomic data have again suggested interbreeding between this species and our own (Peyrégne et al., 2024). Nonetheless, comparisons between genomes have established that the human genome contains numerous mutations that are not shared with Neanderthals and Denisovans, and these mutations are particularly represented in genes involved in neural development and function (Kuhlwilm and Boeckx, 2019; Schaefer et al., 2021), which suggest that brain evolution took humans along a unique path.

The oldest evidence of *Homo sapiens* in the fossil record dates to around 300,000 years BP, although these early specimens do not exhibit the same cranium structure as anatomically modern humans (Harvati and Reyes-Centeno, 2022). Some of the genetic differences between our modern species and chimpanzees emerged before this early time point and are shared with other archaic *Homo* species. For instance, the human variant of the *FOXP2* gene is also found in Neanderthal genomes, which suggests that this variant underwent selection more than half a million years ago. The *FOXP2* gene is involved in vocal learning and underpins human speech production, which is a key aspect of human culture. Selection in the human lineage after the split with archaic *Homo* species did not impact the *FOXP2* gene itself but did influence genes that regulate the expression level of *FOXP2* (Atkinson et al., 2018; Maricic et al., 2013). Our ability to learn socially is likely to have influenced selection on a range of regulatory genes underpinning traits such as language (Varki et al., 2008) and might even explain the apparent acceleration of genetic evolution in our species (Zhen et al., 2021). Differences between human genomes and those of chimpanzees and archaic *Homo* species could thus have resulted from our larger brains and our culture.

Yet, in more recent time periods, has human culture changed too quickly to have impacted brain evolution? The lactose persistence case study demonstrated that genes involved in digestion have undergone recent selection, but could the same be true for genes involved in brain function? Evolutionary Psychologists have argued that human brains are co-adapted gene complexes that are unable to respond quickly to selection (e.g. Buss, 2008; Tooby and Cosmides, 1990a, 2005). These researchers suggest that brains are far too complex for mutations to be anything other than highly detrimental and that the fine balance of the nervous system is incompatible with rapid genetic evolution. In fact, although there is evidence that, within primates, brain tissue has evolved more slowly than that of other organs, accelerated evolution of brain tissue has been specifically found within the human lineage (Khaitovich et al., 2006). As discussed below, evidence has also been found for recent selective sweeps in the human genome during the past 50,000 years, including numerous genes that are involved in brain function. In addition, most of the differences between human genomes and those of other species involve changes in regulatory genes that influence the expression levels of other genes (Liu et al., 2021), which means that human-specific genetic variants might have relatively subtle effects on brain function. Changes in the efficiency of a neurotransmitter receptor, the amount of an activating substance, or the levels of a transporter molecule could potentially produce behavioural and cognitive changes without dramatically disrupting brain function. Thus, the idea that human brains are too complex to have undergone recent rapid selection, as suggested by some Evolutionary Psychologists, is not supported by the current evidence.

Genes that exhibit signs of positive selection in humans include those involved in the expression of glutamate and glycine receptors, olfactory receptors, serotonin transporters, synapse-associated proteins, and neuronal signalling (López Herráez et al., 2009; Voight et al., 2006; Wang et al., 2006; Williamson et al., 2007). Of course, not all genetic variation will have arisen through selection, and not all will have enhanced biological fitness. Some genes that have been subject to recent selection have been linked to disorders such as autism, attention-deficit/hyperactivity disorder (ADHD) and schizophrenia. For instance, some dopamine receptor *DRD4* polymorphisms carry an increased risk to individuals of developing ADHD. These findings have led to the suggestion that rapid brain evolution has left humans particularly susceptible to neurodevelopmental disorders and neurodegenerative diseases (Zug and Uller, 2022). Researchers have begun to apply Gene–Culture Coevolutionary frameworks to the study of cross-cultural variation in psychological traits, such as personality and social norms (e.g. Lee et al., 2022). However, caution is required when interpreting the empirical data. The genomic methods that are deployed detect correlations between genetic variants and phenotypic traits, not causal relationships. Most human traits are influenced by hundreds of genes, as well as a multitude of other factors, so the effects of any given gene variant is typically miniscule. In addition, the methods used to detect signals of recent selection typically focus on those areas of the human genome that show variation across populations, for instance, in terms of the number of repeats of a particular gene (Gao, 2024). For this reason, genetic variants that have been identified as showing evidence of recent selection may be biased towards those not yet fixed across all members of our species.

Most importantly, given that the idea that genetic differences potentially underpin variation in behavioural or cognitive traits is fraught with echoes of scientific racism, the genetic and phenotypic variation in such studies does not segregate on racial grounds (i.e. genetic differences arise between individuals or populations, but *not* between races). In fact, numerous lines of evidence are inconsistent with racist views. For instance, humans across the globe are unusually closely related to each other, due to bottlenecks in ancestral population sizes. Indeed, the amount of genetic variation between humans living on different continents is substantially smaller than the genetic variation found between groups of chimpanzees (Bowden et al., 2012). For comparison, fewer than one in a thousand nucleotide bases differ between any pair of humans, whereas human and chimpanzee genomes differ by fifty times this amount. These data imply that gene–culture coevolution is far more likely to have favoured universal human characteristics than traits that vary between populations. In addition, genetic variation across human populations is *clinal*, meaning that traits change gradually with geographical distances rather than showing distinct boundaries (Maglo et al., 2016). If, as reported, up to 10% of the human genome has been subject to recent selection, it follows that recent selection has not impacted more than 90% of the genome, where drift and serial founder effects are thought more important. Thus, the distribution and evolution of genes for skin colour and perhaps some other visible traits, which were subject to recent selection, are actually not really representative of the rest of the genome. For all these reasons, neither human genomic data nor Gene–Culture Coevolution studies are consistent with a biological concept of 'race' (Benn Torres, 2020; Henrich, 2016; Hochman, 2021).

Culture and the evolution of 'cognitive gadgets'

Compared to non-human animals, human beings have unusually potent capabilities across a range of cognitive traits, such as problem-solving, social cognition, mental time-travel, tool use, and communication (Laland and Seed, 2021). Yet, the human brain shares many commonalities with the brains of our closest living relatives and recently extinct ones. Human brains are three times the size of the brains of other primates, and exhibit high levels of interconnectivity and plasticity, but the basic structure is not fundamentally dissimilar (Herculano-Houzel, 2020). Can these features of the human brain explain the vast differences in cognitive abilities between ourselves and other animals? How does our capacity to generate and transmit cultural knowledge arise, and to what extent does an understanding of human evolution require a consideration of culture? Gene–Culture Coevolution researchers contend that, as the amount of cultural niche construction that our species engages in far surpasses that of any other species, human evolution can only be understood by examining how genes and culture have coevolved. Other researchers, by contrast, place emphasis on genetic or cultural evolution alone, either arguing that culture can be explained as an emergent property of our evolved psychological mechanisms, or alternatively that our ability to engage in social learning is enough to have initiated a ratcheting effect on cognition (Baumard et al., 2024; Birch and Heyes, 2021; Heyes, 2018). The key question here is which approach provides the most complete understanding.

Evolutionary Psychologists have contended that human cultural variation can be explained through two processes: firstly, some cultural variation is 'evoked' by specific cues in the environment (e.g. a particular foraging method or marriage system might be exhibited because of relevant aspects of the local environment); and secondly, much cultural variation is 'transmitted' only in a relatively ephemeral way, without being dependent upon specific environmental cues, and can be thought of as 'epidemiological noise' (Del Giudice et al., 2015; Gangestad et al., 2006; Nettle, 2009; Tooby and Cosmides, 1992). In both cases, culture is not considered to play a causal role in the evolutionary process. These researchers suggest that cultural artefacts, such as clothes, buildings, and laws, can be thought of as 'extended phenotypes', just like the dams, webs, and mounds of non-human animals (Baumard et al., 2024). Cultural phenomena are thus considered to be the product of our genes and reflect the flexible cognitive mechanisms that we possess as a result of past selection. This viewpoint presents a unidirectional causal arrow going from genes to culture and relegates culture to an emergent property of gene expression, albeit one that exhibits variability and environmental sensitivity. Gene–Culture Coevolutionists instead state that ignoring the bidirectional interactions between genes and culture provides an incomplete understanding of human evolution (Boyd and Richerson, 1985; Creanza and Feldman, 2016; Odling-Smee et al., 2003). For example, culture can potentially explain the rapid expansion of the human brain and other aspects of our morphology and life histories (Markov and Markov, 2020; Muthukrishna et al., 2018). Moreover, cultural artefacts, such as weapons, buildings, or laws are not, strictly speaking, 'extended phenotypes', as they are not biological adaptations; they possess design features that arose through cultural evolution rather than through genetic change.

Another presumed challenge to the Gene–Culture Coevolutionary approach has come from the idea that our ability to engage in social learning, along with other very general forms of learning, can explain the complexity of human culture (Birch and Heyes, 2021; Heyes, 2012). The bidirectional interaction between genes and culture is suggested to have been overstated, as a large proportion of what is unique about our species, including our cumulative culture, can be explained without a need to infer genetic changes over recent evolutionary timescales. Theoretical psychologist Cecilia Heyes (2018) from the University of Oxford, UK, has called these emergent psychological processes 'cognitive gadgets'. Heyes argues that the capacities for language, imitation, social cognition, and meta-cognition (the ability to think about one's thought processes) are 'cognitive gadgets' built during the lifespan of an individual via basic processes such as high-fidelity copying behaviour and conditioning (e.g. Heyes et al., 2020). An example is the human ability to read. Reading has its own distinctive neural circuitry, and brain imaging studies have shown that learning to read has major reorganizational effects on the brain. Yet writing was only invented around 5,000–6,000 years ago, which is perhaps too recent for the genetic evolution of reading mechanisms. Thus, the ability to engage in culture, for example, through reading, is at least partly unpinned by culture itself.

At the same time, Gene–Culture Coevolution researchers would suggest that it is difficult to conceive of human cognition and cultural evolution being completely divorced from genetic evolution. As described above, genetic studies have identified hundreds of human genes subject to recent positive selection that show major changes in expression in the brain. Is it plausible that all such changes are unconnected to enhanced cognition or cultural learning? Studies show that much evolution has involved the re-use of pre-existing developmental mechanisms. Reading may be able to develop as a cognitive gadget because it is able to 'reuse' pre-existing genetically evolved circuitry (e.g. for speech). Thus, the suggestion that reading is a 'cognitive gadget' need not imply that our language capability, or cognition in general, is comprised solely, or primarily, of such phenomena (Laland and Seed, 2021).

Granted, the mechanisms that underlie our species' capability for cultural evolution are deployed in a flexible manner that varies between individuals in relation to age, context, and culture, and, of course, children are not born with fully developed sets of cognitive traits, such as mind-reading, or social norms. Yet, the flexible deployment of cultural learning does not constitute compelling evidence for an exclusively cultural evolutionary aetiology. Humans may possess a uniquely powerful cognition, in part, because extensive reuse of neural circuitry has generated greater neural integration. Heyes (2018) asserts that genetic evolution has played only a limited role in the history of our species, conferring features such as a larger brain, longer childhood, and more powerful domain-general cognition (e.g. enhanced working memory and inhibitory control). However, there is now extensive evidence that these features evolved in response to human cultural activities (Henrich, 2016; Laland, 2017), again implying reciprocal causation. The argument therefore comes down to whether weak or strong forms of gene–culture coevolution best explain the evolution of specific traits, which is a matter for future theoretical and empirical research to resolve.

Conclusion

The sub-field of Gene–Culture Coevolution is moving towards a coalescence of multiple fields into a genuinely transdisciplinary science. For instance, geneticists and evolutionary biologists seeking to understand temporal and spatial variation in human allele frequencies, or human uniqueness, will now need to consider the role of cultural variables. These researchers may need to work side by side with anthropologists, archaeologists, or theoreticians, in order to establish whether the molecular signatures of selection they observe are indeed the result of gene–culture interactions. While non-human animals that exhibit extensive social learning and local traditions are likely to experience gene–culture coevolution (Whitehead et al., 2019), the interactions between these two inheritance systems are expected to be particularly pertinent for our species, given the extensive nature of our cultural niche construction and the regulation of our social life by culturally transmitted norms and institutions. Gene–Culture Coevolutionary analyses suggest that evolution in species with a dynamic, socially transmitted culture may be different from evolution in other species, for at least three reasons. First, culture is a particularly effective means of modifying natural selection pressures and driving the population's biological evolution, as was the case for lactase persistence. In addition, by affecting the movements of peoples and fuelling population growth, culture is likely to have had strong and, thus far, comparatively unexplored impacts on other evolutionary processes, such as genetic drift and gene flow. Secondly, culture may generate new evolutionary processes, for instance, cultural group selection. Thirdly, cultural transmission may strongly affect evolutionary rates, sometimes speeding them up and sometimes slowing them down. Such findings suggest that traditional evolutionary approaches to the study of human behaviour may not always be adequate.

8
Comparing and integrating approaches

We now return to the question of to what extent evolutionary theory can help us to understand human behaviour and society. We have seen that the history of taking an evolutionary perspective on human behaviour has been filled with misdemeanours that cannot be ignored. Yet there is also compelling evidence that the careful use of evolutionary theory can increase our understanding of humanity. The preceding chapters have established that there are numerous ways to exploit evolutionary theory to investigate human behaviour, each of which has provided valuable and novel insights. While it is common for each school of thought to portray its methods and reasoning as the way forward, in truth each approach has its strengths and weaknesses. But how do they all fit together?

Our aim in this final chapter is to compare the four contemporary approaches (Human Behavioural Ecology, Evolutionary Psychology, Cultural Evolution, and Gene–Culture Coevolution), to discuss to what extent they are complementary or conflicting, and to explore what each can contribute to the complete picture. However, having delineated research into human behaviour and evolution into particular sub-fields, it is only appropriate that we stress that the field is not quite as easy to partition in real life. There is much common ground and considerable overlap in perspective and methodology between the four contemporary approaches, and individual researchers might well feel that their research draws on multiple approaches and cannot easily be subsumed under only one heading. We accept that the exercise in which we have engaged (i.e. attempting to crystallize distinct clusters of views) will inevitably create the impression that the boundaries between these schools of thought are cleaner than they actually are. Far from wishing to establish or reinforce artificial frontiers, we would like to encourage the evolutionary minded to move selectively between schools, picking and choosing the best tools available, drawing insights from each, and synthesizing divergent perspectives in a critical and discerning manner. The contemporary evolutionist is not tied to any one of the approaches and is free to draw from the methods available across the broader field of the evolutionary behavioural sciences.

We endorse attempts at integration and believe that evolution enthusiasts are more likely to be successful if they are aware of the merits and limitations of each approach.

This chapter begins with a reminder of what we have covered so far, in particular providing a recap of the strengths and weakness of each sub-field of the human evolutionary behavioural sciences. We then briefly describe the role that comparisons with other animal species can play in expanding our understanding of human behaviour and cognition. Focusing on the example of infanticide, we show how a single topic can be investigated separately from each perspective, collectively generating answers that address all of Tinbergen's four questions. We then discuss how the integration of different approaches may provide the broadest explanation of human behaviour, using the study of human foraging skills as an example. Finally, in the last section, we evaluate whether the contemporary sub-fields are complementary, or whether there are some remaining differences and points of conflict, focusing on the issues of the role of culture in the evolutionary process and whether the concept of 'human nature' is useful. We conclude by describing some recent advances in evolutionary theory, which have led to the proposal of an 'extended evolutionary synthesis', and we provide our thoughts on how, by taking an extended evolutionary approach, future research might provide a more complete and nuanced evolutionary perspective on human behaviour.

Summary so far

Let us briefly remind ourselves of what we have covered so far in this book. In Chapter 1, we described our aims and our commitment to providing a balanced approach when describing and evaluating this research field. Although our own research aligns more with some of the sub-fields than others, we have attempted throughout to present accurate descriptions of differing views and provide fair, constructive evaluation. In Chapters 2 and 3, we described the history of applying evolution to human behaviour, starting with the work of Charles Darwin and finishing with the 'sociobiology debates'; this history provides the context for why the social science disciplines largely rejected the idea that evolutionary theory might shed light on human behaviour. The scepticism within social sciences remains to the present day; for instance, many social science textbooks quote the work of human sociobiologists and immediately dismiss such research as provocative storytelling that relies on genetic determinism and bolsters prejudiced political views (Takács, 2018).

Such textbooks tend not to acknowledge the differing perspectives on how to apply evolutionary theory to human behaviour that have emerged since the 1970s. We strongly believe that understanding the history of this research field provides a crucial backdrop for evaluating contemporary approaches and for learning from the past, and students are particularly encouraged to keep this history in mind. In Chapter 3, we also introduced the idea of the 'meme' as a cultural variant that spreads by virtue of being attractive or compelling to the human mind. In this section, we will apply a 'meme's-eye view' reasoning to consider which of the sub-fields has been the most successful in terms of popularity and effectiveness at garnering new recruits, in addition to reminding readers of the pros and cons of each of the four contemporary sub-fields. We will also consider whether certain features of how science is conducted in the sub-fields might have limited or strengthened their success.

In Chapter 4, we described Human Behavioural Ecology, which lays emphasis on the flexibility and adaptability of human behaviour. This research has mostly been conducted in small-scale societies using standard anthropological fieldwork techniques, along with demographic and historiographic methods, to examine whether the expectation that behaviour is *optimal* provides a good fit to the observed data. These researchers are interested in adaptive tradeoffs between different aspects of human life history, as well as the idea that optimality models can provide an understanding of the decisions and priorities that humans make when allocating their time and energy. This research is generally built on high-quality datasets, including from longitudinal studies, which can be challenging due to the difficulties of maintaining and funding such research (McElreath and Koster, 2024). These practical challenges perhaps partly explain why Human Behavioural Ecology remains a relatively small field in terms of numbers of researchers engaged in this approach. The criticism that Human Behavioural Ecology has played down the idea that some aspects of human behaviour might be maladaptive is perhaps misplaced. Using optimality as a starting premise for modelling human behaviour does not mean that researchers are unaware that sometimes their models will not provide a good fit for the data, and iterative revisions of a model can often provide a greater understanding of underlying processes. The fact that human beings frequently construct conditions that are suited to them also means that behaviour might be adaptive much of the time. We examined the benefits of taking a piecemeal approach when studying human behaviour and showed that Human Behavioural Ecology has provided novel insights in the behaviour and demography of contemporary and historical human populations (Gibson and Lawson, 2015; Koster et al., 2024; Sear, 2015).

In Chapter 5, we presented the sub-field of Evolutionary Psychology, defined by its focus on evolved psychological mechanisms and the assumption that such mechanisms reflect selection that acted on our ancestors. This sub-field has been built on the assumption that the human mind contains psychological mechanisms that provide solutions to recurring problems faced by our ancestors, along with the expectation that modern, post-industrial environments are sufficiently different from those of our ancestors to have generated 'mismatches'. Evolutionary Psychology has focused on universal features of the human mind that are shared by all members of our species or by specific subgroups, such as men and women. Whether enough is known about where and how our ancestors lived in the Pleistocene to make specific predictions about human cognition remains a challenge for this sub-field. Evolutionary Psychology has encompassed the idea that the expression of evolved psychological traits is sensitive to cues across the lifespan. In this way, Evolutionary Psychologists are perhaps studying the psychological mechanisms that might give rise to the diversity of behavioural outcomes that are measured by Human Behavioural Ecologists (Kaplan and Gangestad, 2005; Sng et al., 2018); we return to this point later in the chapter. Evolutionary Psychology has certainly been successful in terms of the large number of researchers using this approach and the volume of research generated (e.g. Buss, 2015; Shackelford, 2020). The popularity of Evolutionary Psychology perhaps partly reflects the ease with which empirical data can be collected, for instance, through relatively quick questionnaire studies. Although Evolutionary Psychology has certainly been impactful, this sub-field has continued to attract criticism (e.g. Smith, 2020), such as in terms of its commitment to a gene-centred view of evolution, as discussed later in this chapter.

In Chapter 6, we introduced the sub-field of Cultural Evolution. The basic idea behind this sub-field is that evolutionary methods developed for studying changes in gene frequency over time might be applied to the study of cultural variants. Although there are some differences in how cultural and genetic inheritance work, Cultural Evolution researchers are well aware of such differences and are able to adapt their models to account for them. These researchers study the biases that influence the transmission of culture within populations, including *direct* biases that depend upon the content of the cultural information and *indirect* biases (such as 'conformity') that depend upon who else in the population has adopted a particular trait. Cultural Evolution research has seen a substantial growth in volume since the turn of the century (Matzig et al., 2023), reflecting a vibrant empirical research programme underpinned by sophisticated mathematical modelling (Tehrani et al., 2024). Critics have suggested that Cultural Evolution research has neglected the idea that minds 'select' culture and that evolved structure in the brain might bias individuals to 'reconstructing' the same information. Although Cultural Evolution has tended to focus on the selection of transmitted information, the idea that psychological processes can bias transmission has been a core part of the sub-field from its inception. Neither the suggestion that minds select culture, nor the dissimilarities between genetic and cultural evolution, are inherently problematic for the sub-field, although a greater understanding of the psychological processes behind cultural transmission and transformation might be warranted.

In Chapter 7, we described the key concepts underlying the sub-field of Gene–Culture Coevolution, including that, because genetic evolution can sometimes be faster than we might expect and cultural evolution can sometimes be slow, genes and culture operate on overlapping timescales. The opportunity for genes and culture to influence each other is thought to underlie the observations from genetics research that recent selective sweeps have taken place in the human genome in response to human cultural activities. Additional empirical evidence has been provided by comparisons with the genomes of our closest living and extinct relatives, as well as archaeological data. We introduced the concept of *niche construction* and the argument that gene–culture coevolution might have played a substantial role in the evolution of our species. We critically evaluated the idea that dissecting culture into discrete packages provides an overly simplistic view of human culture, while also pointing out the merits of taking a pragmatic approach to quantifying culture. The question of whether the human brain is too complex to have been influenced by gene–culture coevolutionary processes was evaluated and shown to be inconsistent with much empirical research. At the same time, caution is required when interpreting genetic data, and simplistic storytelling must be avoided, particularly bearing in mind the spectre of scientific racism. Proponents of Gene–Culture Coevolution have argued that this sub-field provides the most comprehensive understanding how the human mind and behaviour has coevolved with our traits (Henrich, 2016; Laland, 2017; Muthukrishna, 2023). The strength of this argument will depend upon the accumulation of empirical data drawn from numerous disciplines, which makes such research challenging.

Cross-species comparisons

Before we take a look at whether the contemporary sub-fields might be combined to provide a more complete understanding of human behaviour and cognition, and examine any barriers to this integration, we return to the question of whether comparisons with non-human animal species have contributed to this effort. As discussed in Chapter 3, human sociobiologists regularly drew on comparisons with non-human animals when interpreting human behaviour. Unfortunately, such comparisons were often arbitrary and speculative, and the conclusions that were drawn were often subject to criticism (Rose et al., 1984). Perhaps as a result, the comparative approach has not become central to any of the contemporary sub-fields, but rather has generated a distinct sub-field of its own, sometimes labelled 'comparative cognition' (Call et al., 2017). Since the time of the sociobiology debates, the amount of empirical research on the cognitive abilities of non-human animals has grown substantially (Kaufman et al., 2021), providing a stronger foundation for such comparisons. In addition, rigorous statistical methods are now available for studying the evolutionary history of cognitive and behavioural traits (Nunn, 2011; Revell and Harmon, 2022). These developments have provided new insights into the cognitive similarities between humans and other species, as well as the features of cognition that might be uniquely sophisticated in our own species. Here, we briefly summarize some of the findings from this comparative work.

At first glance, our species is unlike any other in terms of our cultural outputs, such as technologies, institutions, and arts, and our domination of the environment. However, claims along the lines of 'humans uniquely do X, or possess Y' have generally fallen by the wayside when such cognitive or behavioural traits have been established in other animal species. One example is tool use. Human beings rely heavily on the manufacture and use of tools, yet tool use is also observed in a broad range of animal species, including insects, reptiles, birds, non-human primates, marine mammals, and cephalopods (Biro et al., 2013; Shumaker et al., 2011). Chimpanzees use sticks for extracting termites from nests and leaves for soaking up water or honey, and also use objects when displaying to others or cleaning themselves (Bandini and Harrison, 2020). Some species can also modify tools to increase their efficiency; for instance, Galápagos woodpecker finches strip down twigs and cactus spines before using them to pry insects from crevices (Tebbich and Bshary, 2004). In which case, aside from the amount of tool use that humans exhibit, is there anything different about the cognitive processes that humans use in relation to tools compared to other species? Children, but seemingly not other animals, are able to understand causal properties, such as gravity and support (Ruiz and Santos, 2013), which may in part help to explain why our species excels at problem-solving and using tools flexibly (Laland and Seed, 2021). In addition, children are particularly capable of learning tool use by watching more experienced individuals, which allows them to master complex skills, such as manufacturing baskets, sledges, and canoes, and hunting animals (Lew-Levy et al., 2017; Ruiz and Santos, 2013). Imitation and teaching are found in animals too but are far more extensive in humans (Hoppitt et al., 2008; Wood et al., 2023). Our ability to learn from, and build on, the knowledge of others might underpin our impressive tool use capabilities.

Our superior social learning could reflect better imitative or teaching abilities, or a more general enhancement of executive functioning in the human brain, such as our working memory and ability to inhibit behavioural responses, switch attention, plan, and multi-task (Laland and Seed, 2021). These aspects of human cognition can be found in other animal species, but they are again particularly well developed in humans. An ability to understand the intentions of others, share attention, and work towards collaborative goals can be found in some other species, although at a precursor level compared to humans (Laland and Seed, 2021). Humans share with many animals the ability to communicate about features of the world, and to learn the meaning of symbols or utterances. For instance, monkeys give different types of alarm calls to different classes of predators, which has been described as 'referential' communication (Cäsar and Zuberbühler, 2012), but such cases fall well short of the infinite flexibility of human language. Some species of birds and mammals engage in vocal learning and exhibit local dialects; for example, sperm whales living in the Pacific ocean can be reliably allocated into six 'clans' based on the patterns of their click vocalizations rather than the geographical area in which they live (Rendell and Whitehead, 2003). However, there is little evidence that bird or whale songs convey complex messages. Human language is perhaps the most obvious domain in which our abilities surpass those of other species, allowing us to communicate with others about past and future events, formulate and express complex thoughts and ideas, and generate symbolic meanings.

Gathering data on the behaviour and abilities of a large number of species allows questions to be asked about the evolutionary history of traits, such as whether specific traits coevolved with other traits or are associated with particular environmental, social, or life history parameters. Such comparisons require researchers to take into account the relatedness between species, as two closely related species might share the same characteristics because those characteristics were also found in their common ancestor. Modern phylogenetic statistical techniques allow researchers to examine whether traits or variables co-occur while controlling for the effects of shared ancestry (Nunn, 2011; Revell and Harmon, 2022). An example of multi-species comparative research is provided by the study of primate intelligence and brain size. Numerous hypotheses have been put forward to explain variation in brain size across extant primate species (DeCasien et al., 2022). Some primate species, such as baboons, chimpanzees, and humans, have larger brains than expected for their body size. These species also live in large social groups, which led to the hypothesis that complex social lives selected for increased brain size in these lineages (Dunbar, 1995). Studies of primate datasets have confirmed that, after controlling for the phylogenetic relatedness between species, relative brain size does correlate positively with social group size (Shultz and Dunbar, 2007). However, primate brain size also correlates with many other aspects of behaviour and life histories, including diet type, levels of extractive foraging, social learning, and innovation (DeCasien et al., 2022; Navarrete et al., 2016). Based on these findings, larger relative brain sizes in primates have been proposed to be associated with enhanced cultural (Street et al., 2017) or general intelligence (Burkart et al., 2017).

Comparative studies will continue to provide novel insights into the evolution of human behaviour and cognition. Some limitations and constraints should be borne in mind, though. For instance, simply comparing human beings with one other species, such as chimpanzees, provides limited information about which traits might be ancestral and which traits are derived (i.e. have undergone selection within a specific lineage) (Laland and Brown, 2008). Broadening such comparisons across multiple species, and beyond primates, can provide additional perspectives on the evolution of cognition. For instance, corvids (e.g. rooks and crows) exhibit complex physical cognition, episodic-like memory, cooperation, and tool use ability (Seed et al., 2009), which has led to them being described as 'feathered apes' (Emery, 2004). Researchers can therefore ask whether the evolution of cognition in these large-brained birds provides an example of convergent selection. Numerous animal species exhibit local traditions and cultures (Laland and Galef, 2009), and there is evidence of gene–culture coevolution occurring in non-human species (Whitehead et al., 2019). In summary, cross-species comparisons are a vital tool for the contemporary evolutionist, providing important insights into our evolutionary ancestry, the selective factors that favoured human cognitive capabilities, and the function of many human attributes. However, such cross-species comparisons must be conducted without the intrusion of anthropomorphism (Zuk, 2002) and are best exemplified by research that critically evaluates the comparative evidence at every turn (e.g. Hrdy, 1999, 2009).

Integrating the contemporary sub-fields

In this section, we examine the extent to which the four contemporary sub-fields can be seen as complementary to each other. We first discuss whether the sub-fields differ in their *levels of explanation* and are therefore addressing related phenomena from different angles. We use human infanticide, more specifically infanticide by mothers and unrelated males, as a case study to illustrate how the different approaches might collectively contribute to a broader general understanding. We then examine whether the ways in which each sub-field engages in *hypothesis generation and hypothesis testing* might allow them to be successfully combined to generate novel, integrative research. We use human foraging skills as a case study of how the different sub-fields have constructively used complementary methods to test specific hypotheses. We go on to discuss another topic—fast–slow life histories—as an illustrative example of where caution is sometimes required when studying a topic from different disciplinary perspectives, and where the misapplication of evolutionary ideas has generated negative outcomes.

Levels of explanation

In Chapters 4–7, we introduced the four contemporary sub-fields and described their key assumptions. However, the histories of these sub-fields have also been shown to be highly intertwined. Some Human Behavioural Ecologists and Evolutionary Psychologists have suggested that their two sub-fields are actually highly compatible with each other, as the flexible behavioural outputs observed by Human Behavioural Ecologist might be the outcome of evolved psychological mechanisms that are proposed by Evolutionary Psychologists (e.g. Kaplan and Gangestad, 2005; Sng et al., 2018). Other Human Behavioural Ecologists have demonstrated affinity with the field of Cultural Evolution (e.g. Borgerhoff Mulder et al., 2009) or have pointed out the potential complementarity between Human Behavioural Ecology and niche construction theory (Laland and Brown, 2006; Ready and Price, 2021). Likewise, those who study Cultural Evolution are generally supportive of a Gene–Culture Coevolutionary perspective, given their shared history and the fact that Cultural Evolutionary processes are key component of both sub-fields. Perhaps each of the four sub-fields can provide complementary information about a particular behaviour or trait? In Chapter 2, we described Tinbergen's (1963) four questions about animal behaviour; to recap, when studying behaviour, researchers could be asking about the *mechanisms* (e.g. genes, hormones) that underpin the expression of that behaviour, the *development* of that behaviour across an individual's lifespan, the *function* of the behaviour in terms of enhancing reproductive success, and the *evolutionary history* of that behaviour in the specific lineage of interest. Along these lines, perhaps the sub-fields are providing different levels of explanation when asking about a specific aspect of human behaviour or cognition.

Let's look at an example. While this topic is challenging to contemplate, infanticide is exhibited in many mammalian species (Lukas and Huchard, 2014; Minocher and Sommer, 2016), which suggests that taking an evolutionary approach might increase understanding of this phenomenon in our own species, too. In Chapter 3, we described how Sarah Hrdy, working within the sociobiological tradition, was able to make sense of the curious observation that female langur monkeys will mate with infanticidal males, which is considered an adaptive strategy on the part of females in response to the high turnover of males in the group. Hrdy (1999) also argued that infanticide by mothers is more common in human beings than is infanticide by 'invading' males. Human infants are very costly to raise, in terms of time, energy, and resources, and mothers require the cooperative help of social companions. The ability of a mother to raise a child successfully may therefore be contingent on the amount of social support that she receives (Hrdy, 2009). Here, a comparative perspective, which considers human infanticide in the context of that exhibited by other primates, has shed light on those aspects of this behaviour that are similar to, and different from, those of closely related species.

Human Behavioural Ecologists have studied infanticide as part of a broader interest in parental care. In certain circumstances, mothers may be selected to terminate investment in a young infant as a strategy to allocate limited resources in a manner that maximizes lifetime reproductive success. Natural selection may have favoured mothers, and sometimes fathers or close relatives, who decrease their investment in, or even kill, offspring where the costs of raising the infant are expected to outweigh the benefits, for instance, where the health of the mother will be compromised by attempting to raise the infant. Historical records (1655–1939) from Ditfurt, Germany, indicated that infant mortality was much higher among illegitimate than legitimate children, and that death rates were particularly high when the mother went on to marry a man who was not the father of the child (Voland and Stephan, 2000). Infanticide was seemingly not due to stepfathers, as the deaths occurred prior to remarriage, but instead suggests an adaptive strategy on the part of the mother who might end up having more offspring in the future by making this sacrifice. Unrelated members of social groups might also kill infants, particularly in the absence of a primary caregiver. For example, a study of the Ache, who live in Paraguay, revealed that children who lost their mothers during their first year of life were often killed (Hill and Hurtado, 1996). Human Behavioural Ecologists thus attempt to understand patterns of infanticide by assuming that adaptive strategies underpin human behaviour (Frankenhuis and Amir, 2022).

In Chapter 5, we described Martin Daly and Margo Wilson's (1988) analysis of infanticide from an Evolutionary Psychology perspective, in which they documented a significantly elevated risk to children residing with stepparents. Daly and Wilson assumed that substitute parents might care less for children than natural parents because they are unrelated, drawing on criminal record data to test their hypothesis. In situations where men are uncertain of the paternity of their partner's child, facial resemblance might provide a useful cue of relatedness (Daly and Wilson, 1982). Family members often comment on whether an infant looks like their parents, and such comments are particularly directed towards the nominal biological father (McLain et al., 2000), even though infants do not generally resemble their father more than their mother (DeBruine et al., 2016). In later childhood, perceived or actual facial resemblance between fathers and children has been reported to correlate with the father's levels of investment and emotional closeness (Alvergne et al., 2009, 2010). The propensity of mothers and other individuals to reassure fathers about the physical resemblance between themselves and offspring might have been selected in past environments as a means of increasing paternal investment and decreasing the risk of abandonment or abuse. Evolutionary Psychologists have thus attempted to understand the psychological mechanisms behind behavioural strategies related to childcare and neglect.

A study by Nan Li and colleagues (2000) illustrates the kind of investigation that Cultural Evolutionists have undertaken to address this topic. These researchers developed Cultural Evolution models that predicted how the sex ratio at birth in China might change over time in the face of state-enforced low fertility and a culturally transmitted preference for sons that has been manifested in female-biased, sex-selective abortion. At the time that the researchers generated their models, the birth sex ratio in China was around 1.20 (i.e. 120 males born for every 100 females) (Jiang and Zhang, 2021). Li and colleagues used survey data to measure the strength of cultural transmission of son preference across generations in two Chinese counties that were chosen as high and low projections for the country at large. Plugging these parameters into their models, they predicted that the national-level birth sex ratio in 2020 would range somewhere between 1.10 and 1.34 (i.e. 110 to 134 males per 100 females), depending upon the future strength of son preferences. By 2020, the birth sex ratio in China had indeed dropped to around 1.10 (Jiang and Zhang, 2021), coinciding with a reduction in son preferences and a relaxation of family size restrictions. The sex ratio of third-born offspring remained high (around 1.36), which suggests that a preference for sons still impacted the sex ratios of later-born offspring. The model predictions therefore closely matched the future birth sex ratio values. Gene–Culture Coevolution researchers also generated models predicting that, depending on the rules of cultural transmission, genes favouring either female-skewed or male-skewed birth sex ratios might increase in frequency if families engaged in sex-selective abortion (Kumm et al., 1994). Hence, several types of mathematical models have been used to examine the relationship between cultural transmission and sex-biased investment.

Irrespective of the methodological differences between the sub-fields, there is little that is conflicting or incompatible about these findings. In fact, each investigation reinforces the others, collectively building up a panoramic view of the topic at hand that spans genetic to sociocultural levels of analysis. Thinking in such terms helps to bring out the complementarity of the aforementioned studies, since all are required to generate the complete understanding demanded by Tinbergen. Indeed, a full grasp, both of the causes of infanticide and of the behavioural and cognitive characteristics that humans have evolved to protect their children, requires all of the above approaches and more, including sociology. Here is an advertisement for pluralism in evolutionary perspective. There is no reason for researchers to restrict themselves to a single research technique when, by and large, the different methodologies are highly complementary. Eric Alden Smith (2000) has made the same point:

> Most evolutionary social scientists and biologists would agree that complete evolutionary explanations of behavior will include (i) heritable information that helps build (ii) psychological mechanisms, which in turn produce (iii) behavioural responses to (iv) environmental stimuli, resulting in (v) fitness effects. (p. 35)

Smith noted that Evolutionary Psychology focuses on (ii) psychological mechanisms and their links to (iii) behaviour and (iv) the environment, whereas Human Behavioural Ecology focuses on (iii) behavioural responses with attention to (iv) environmental stimuli and (v) fitness effects. Cultural Evolution is concerned with (i) the cultural component of heritable information, which is expressed in (iii) behaviour, and Gene–Culture Coevolution focuses on (i) both genetic and cultural inheritance, and their links to (ii) psychological mechanisms and (v) reproductive success. We agree with Smith (2000, p. 35) that, 'Viewed in this light, a tentative case can be made for explanatory complementarity'.

Hypothesis generation and hypothesis testing

As discussed in the previous chapters, each of the contemporary sub-fields has an independent, though intertwined, intellectual history, with each sub-field having its roots within specific disciplinary contexts. For example, many researchers who align themselves with Human Behavioural Ecology are trained as anthropologists, whereas many Evolutionary Psychologist researchers have a background in psychology. Cultural Evolution research was initially dominated by mathematical modelling approaches, thus attracting researchers with strong mathematical skills; mathematical approaches remain a key focus of this sub-field, alongside empirical research in laboratory and field settings, which has drawn in psychologists. Gene–Culture Coevolution research also has a strong mathematical component and has attracted multidisciplinary teams that often include biological anthropologists and geneticists. These distinctions between sub-fields are reflected in the manner in which researchers generate and test their hypotheses. We have summarized some of the key difference between sub-fields in Table 8.1, including the 'level of explanation' as described in the previous section. Here, we summarize differences in hypothesis generation and hypothesis testing between the sub-fields and examine whether integrative research is possible.

To generate their hypotheses, Human Behavioural Ecologists start with the premise that human behaviour can be understood using an optimality approach, and their predictions about behaviour are often based on formal optimality models derived from the animal behavioural ecology literature. The predictions of these models are then compared to empirical data, most commonly ethnographic data collected in small-scale societies or data from historical records. In contrast, Evolutionary Psychologists often use broader interpretations of evolutionary theory, such as inferences about how selection might have acted on ancestral populations (i.e. without comparison to the predictions of specific models) to generate hypotheses about the evolved psychological mechanisms that underpin human behaviour. Empirical data are generated by Evolutionary Psychologists using multiple methods, with a particular focus on standard psychology techniques such as questionnaires and laboratory experiments, or data are extracted from archival records. Cultural Evolution researchers draw on analogies with genetic evolution and deploy mathematical models to generate hypotheses about how people's behaviour, beliefs, and institutions are shaped by socially transmitted information. Such models are used to simulate the transmission of cultural information

Table 8.1 Comparing the five approaches

	Human Behavioural Ecology	Evolutionary Psychology	Cultural Evolution	Gene–Culture Coevolution
Level of explanation	Behaviour	Psychological mechanisms	Cultural traits	Gene–culture combinations
Hypothesis generation	Optimality models	Inference from evolutionary theory or Pleistocene ancestors	Mathematical models	Mathematical models, genetic data
Hypothesis-testing methods	Quantitative ethnographic information, historical records	Multiple, but commonly questionnaires, lab experiments, historical records	Mathematical modelling and simulation, laboratory experiments	Mathematical modelling and simulation
Comparator	Animal behavioural ecology	Pleistocene hominids	Genetic evolution	None
Is behaviour adaptive?	Yes, as a starting assumption	Not always, because of adaptive lag	Not always, because of Cultural Evolution	Often, but not always because of Cultural Evolution
What is culture?	Multiple descriptions, but mainly behaviour elicited by ecological conditions	Multiple descriptions, but mainly cultural universals constrained by evolved psychological mechanisms	Socially transmitted information guided by learning biases	Socially transmitted information guided by learning biases
What are humans?	Sophisticated animals characterized by extreme adaptability	Sophisticated animals guided by psychological adaptations	Sophisticated animals shaped by context biases and local norms	Sophisticated animals guided by genetic and cultural information

Based on Smith (2000).

and evolution of behaviour, and empirical data generated through laboratory or field studies are compared to the model expectations. Gene–Culture Coevolution researchers similarly construct mathematical models that describe how the two streams of inherited information (i.e. genes and culture) interact to generate human behaviour. The types of data that inform these models include genetic data, archaeological data, and information about the cultural evolution of traits. All of the sub-fields engage in an iterative process of knowledge production, whereby the generation of new data and insights can lead to the refinement of underlying assumptions about the variables that influence behaviour.

An example of how the methodological approaches of two sub-fields, in this instance Human Behavioural Ecology and Cultural Evolution, are complementary is provided by research on the social learning of foraging skills. Human Behavioural Ecologists have collected long-term data on how children learn the skills required for hunting and gathering in a number of small-scale societies, providing insights into the cultural transmission of such skills within and between generations. For example, among the Tsimane farmer-foragers in Bolivia, young children learn foraging skills from related adults, particularly mothers, whereas those in middle childhood and adolescence switch to co-foraging with peers (Jang et al., 2024). Among the BaYaka in the Congo Basin, the transmission of foraging skills is particularly strong between mothers and daughters, and between fathers and sons, generating a gender-based structure to learning (Schniter et al., 2023). Using parameters from such studies, Cultural Evolutionists have modelled how population structure might influence cultural transmission, leading to testable predictions. For instance, the transition from learning from older to younger individuals is predicted to occur sooner when rates of environmental change are higher and when subsistence patterns reduce parent–offspring contact (Deffner and McElreath, 2022; Fogarty et al., 2019), and the division of knowledge into particular demographic groups is predicted to increase the risk of foraging innovations being lost (Ben-Oren et al., 2023). The incorporation of population structure into Cultural Evolution models, prompted by detailed ethnographic work, demonstrates the benefits of engagement between the sub-fields.

As an example of how care must be taken when generating hypotheses, we consider the concept of 'fast–slow' life histories. Within evolutionary biology, this concept relates to the observation that some species are characterized by a fast progression to reproductive maturity, large numbers of offspring, and a short lifespan, whereas other species have longer periods of maturation, fewer and more costly offspring, and longer lifespans. These concepts have subsequently been applied in psychology to within-species variation in human behavioural and cognitive traits. However, a number of issues immediately arise. Firstly, the concept of fast–slow life histories in evolutionary biology relates to comparisons between, not within, species; between-individual differences are more appropriately framed in terms of phenotypic plasticity (Zietsch and Sidari, 2020). Secondly, where the concept of fast–slow life histories has been applied within species, the current empirical data from birds and mammals, including humans, reveals that individuals do not simply fall along a continuum from 'fast' to 'slow' (Van de Walle et al., 2023). Within-species variation is better described by other dimensions that vary across species. Thirdly, life history theory generates predictions about characteristics such as maturation rates and reproductive schedules, not behavioural and cognitive traits (Sear, 2020). The life history literatures in psychology and evolutionary biology have developed with relatively little overlap (Nettle and Frankenhuis, 2019), and, disturbingly, the former has been used to bolster eugenic ideologies (Graves and Goodman, 2021; Sear, 2021a). This example therefore highlights the importance of gaining a deep understanding of the relevant evolutionary biology literature when generating hypotheses about human behaviour and cognition.

Two other areas where the contemporary sub-fields differ in their approaches to generating and testing hypotheses relate to the comparator that they use, and whether the research starts with the assumption that human behaviour is currently adaptive. The comparator refers to the entity with which comparisons are made in order to generate explanatory hypotheses, and, as shown in Table 8.1, the sub-fields differ in the types of comparators that they use. In terms of the question of whether behaviour is adaptive, Human Behavioural Ecologists start with the assumption that human behaviour is adaptive (i.e. it varies in response to current conditions so as to enhance reproductive success), while Evolutionary Psychologists primarily endeavour to isolate human adaptations (i.e. characters favoured by natural selection for their effectiveness in a particular role). Not only are adaptations and adaptive behaviour not the same, but they can be regarded as orthogonal or independent, as discussed in Chapter 4. Both approaches are necessary if the characteristic is to be fully understood as a product of evolution, with its function and history each well understood. At the methodological level then, the approaches exploit alternative forms of data collection and hypothesis testing, and there are considerable differences of focus and emphasis. Rather than reiterating these points, we now turn to the question of whether there are fundamental differences between the sub-fields that are likely to create barriers to integration.

Points of contention, and future directions

In this final section, we discuss three specific topics where we believe that important points of contention, or areas for further integration, remain. First, we focus on the question of the role of culture in the evolutionary process, given that some sub-fields describe culture as solely the product of past selection on genes, whereas other sub-fields place culture centrally in the evolution of our species. Second, we discuss the question of whether 'human nature' is a useful concept or whether it potentially constrains progress in understanding human behaviour and cognition. Finally, we discuss some recent developments in evolutionary theory, specifically the emergence of the *extended evolutionary synthesis*, and discuss the potential implications of this broader evolutionary framework for future research on human evolution.

What is culture?

A recurring theme within this book is conveyed by the questions '*What is culture?*' and '*What role does culture play in the evolution of our species?*' As shown in Table 8.1, the sub-fields vary in their descriptions and explanations of human culture. To briefly recap, Human Behavioural Ecologists view cultural transmission as one of the mechanisms by which human beings might exhibit adaptive behaviour, and sometimes maladaptive behaviour, in response to cues from the physical and social environment (Mace, 2014). As the main focus of Human Behavioural Ecology is whether the concept of optimality can shed light on behavioural diversity, these researchers often remain agnostic about the proximate mechanisms that might underlie such diversity (the *phenotypic gambit*). However, as demonstrated by the above example of social learning of foraging skills, Human Behavioural Ecologists are increasingly considering the role of culture in their empirical work and in their models (e.g. Borgerhoff Mulder et al., 2009). Human Behavioural Ecologists have also addressed questions about the developmental and hormonal mechanisms underpinning human behaviour and life histories (e.g. Bribiescas, 2016; Jasienska, 2013). A rigid commitment to the phenotypic gambit therefore appears to be waning in this sub-field and in evolutionary demography more generally (Sear et al., 2016), which is likely to increase the predictive power of these models (Borgerhoff Mulder, 2013; Brown, 2013). Human Behavioural Ecologists are also increasingly applying their methods to populations that have transitioned away from small-scale subsistence methods and undergone demographic transitions to low fertility (Mattison and Sear, 2016). Studying cultural change is therefore not inherently antithetical to the Human Behavioural Ecology approach, and cultural diversity is the core component of human behaviour that this sub-field endeavours to understand.

Evolutionary Psychology has focused on those aspects of human behaviour that are assumed to be exhibited across all populations or that characterize major subgroups, such as men and women. Cultural variation is considered as either the outcome of universal evolved psychological mechanisms that generate different outcomes in different settings or ephemeral cultural changes that have little impact on how our species has evolved (Nettle, 2009; Tooby and Cosmides, 1992), and some but perhaps not all Human Behavioural Ecologists concur with this position. From this perspective, culture is a 'proximate' rather than an 'ultimate' or evolutionary cause (c.f. Mayr, 1961). Evolutionary Psychologists consider *context-dependent strategies*, whereby local environmental variables might shift the behavioural output of universal psychological mechanisms in an adaptive manner, and also consider how the expression of evolved psychological traits might vary across individual lifespans (Bjorklund and Pellegrini, 2000; Gangestad and Simpson, 2000; Tooby and Cosmides, 1992). Cultural variation is thus seen as the product of flexible, evolved structure in the brain that humans possess as a result of past selection. The hypothesis that a 'mismatch' occurs between these psychological mechanisms and current environments remains a prevalent assumption (e.g. Goetz et al., 2019; Li et al., 2018). For these researchers, making inferences about how natural selection has acted in past environments is deemed the appropriate approach to understanding the evolution of culture-generating human brains (Baumard et al., 2024). As in the broader field of psychology, Evolutionary Psychology research has mainly been conducted on participants from Western, educated, industrialized, rich, and democratic (WEIRD) societies (sensu Henrich et al., 2010). However, calls for diversifying the sub-field beyond such populations are leading to greater engagement in cross-cultural research (Apicella et al., 2020), which is likely to generate a fuller understanding of the human mind given that behavioural diversity is a characteristic feature of our species.

Cultural Evolution and Gene–Culture Coevolution researchers view culture as socially transmitted information that is guided by evolved learning rules and motivational priorities (Richerson and Boyd, 2005). Culture is seen as a dynamic evolutionary system in its own right, and the content of socially acquired knowledge underpinning human culture is not considered to be overly pre-specified by evolved structure in the brain. That is, culture is viewed as an 'ultimate', or evolutionary, cause rather than solely a 'proximate' cause, while genetic evolution is assumed primarily to specify *how* to learn rather than *what* to learn. Cultural Evolution researchers propose that social learning biases, such as 'copy the majority', minimize the constraints on what humans learn and can potentially result in the spread of cultural traits that are maladaptive in terms of genetic fitness (Kendal et al., 2018). Therefore, in contrast to the functionally specialized psychological mechanisms that have been emphasized by Evolutionary Psychologists, human culture, and human behaviour in general, is considered by Cultural Evolutionists to be largely underpinned by domain-general learning and cognition.

Here then is a genuine difference between the sub-fields that can be answered through continued empirical research: '*What is the relative importance of content versus context biases (or domain-specific versus domain-general mechanisms) in shaping human cultural learning?*' and '*What role has, and does, culture play in human evolution?*' The overarching proposition from Cultural Evolutionists is that cultural phenomena cannot be fully understood without recourse to the intrinsic processes of cultural change. Cultural Evolution researchers conduct cross-cultural studies and employ phylogenetic comparative methods to understand cultural diversity, as well as devising mathematical methods that allow inferences to be drawn from cross-cultural datasets (e.g. Deffner et al., 2022). Gene–Culture Coevolution researchers hold a similar view of culture to Cultural Evolution researchers, while also emphasizing the bidirectional relationship between genetic and cultural evolutionary processes. Therefore, unlike most Evolutionary Psychologists, these researchers do not agree that culture can be understood simply as the result of past selection on genes and instead argue that culture plays a causal role in the evolutionary process. The answers to the questions outlined at the start of this section remain disputed and represent a key point of contention between the sub-fields. We return to the topic of causality in the final section of this chapter.

The concept of 'human nature'

Here, we examine whether the concept of 'human nature' is a help or a hinderance in understanding the evolution of human behaviour and cognition. A colloquial definition of human nature might be that it describes the general set of characteristics that all human beings share or that set us apart from other species. These features perhaps form the essence of what it means to be human. All of the contemporary sub-fields in the human evolutionary behavioural sciences agree that the human brain has evolved and that all humans share substantial components of their behaviour and cognition. Some researchers have attempted to document universal characteristics that are shown across all human populations, with the initial focus being on traits such as incest avoidance and divisions of labour (Brown, 1991; Wilson, 1994). However, are researchers any closer to cataloguing these traits? Is the concept of human nature limiting our understanding of the traits that it aims to define? Clearly, the explanatory tools available to the evolutionist go far beyond the mere documenting of traits that humans have in common. Rather than attempt to generate our own description of 'human nature', we evaluate how this concept has been defined and implemented.

The founders of Evolutionary Psychology, John Tooby and Leda Cosmides (1990b, 2005), explicitly stated their intention to defend the concept of a universal human nature, following the lead of Edward Wilson (1975a, 1978). Generating a description of human nature might fulfil the same function as providing an entry in a field guide, such that, if an extraterrestrial being encountered a human from anywhere in the world, that human might be reasonably expected to exhibit a particular set of psychological traits or developmental outcomes. Edouard Machery (2008), a philosopher of science, has argued that human nature is a useful concept for describing a typical, though not necessarily universal, set of properties that human beings possess as a result of evolution. Applying this reasoning, human nature would overlap substantially with what might be described as 'primate nature' or even 'vertebrate nature'. As described earlier in this chapter, the question of what makes humans cognitively distinct from other species remains a topic of discussion, given the evidence for some impressive cognitive abilities in other animals. What sets human beings apart from other species is perhaps our highly developed executive functioning, social cognition, tool use, and capacity for culture and language (Barrett, 2020; Laland and Seed, 2021); yet less developed versions of most of these abilities have been documented in other animals.

The idea that 'human nature' might be used to define that which is typically or distinctively human is potentially problematic then, given that we share many psychological traits with other species, even if our capacities are greatly enhanced. Indeed, the features of humanity that are most distinct and maximally diagnostic—language, technology, and culture—are precisely the entities generally associated with 'nurture'. If the goal is to describe what typical human beings are like with respect to their cognitive traits, this comparative work can be conducted without having to place particular traits into, or outside of, a category called 'human nature'. That wider comparative perspective would have the additional advantage of encouraging the study of human diversity, rather than focusing on 'core' features. The fields of cultural psychology and cross-cultural neuroscience are already engaged in the process of documenting the variability of human cognitive traits (Heine, 2020; Kim and Sasaki, 2014), as well as reflecting on how a 'universal' trait might be defined (Norenzayan and Heine, 2005), and similar considerations are likely to benefit the human evolutionary behavioural sciences.

In addition to the field guide approach, the concept of human nature has been used in another way, which is to set apart 'human nature' from 'human culture'. This version of human nature presupposes that some aspects of human behaviour are 'evolved', 'innate', or 'instinctive', whereas other aspects, particularly those that are defined as cultural, are learned, plastic, and variable. In Chapter 2, we described how the idea of human 'instincts' was rejected within psychology with the rise in behaviourism and then reprieved by ethologists studying animal behaviour. The ethologists conceded that 'instinct' is not an adequate explanation for animal behaviour, as it discourages consideration of how behaviour develops (Hinde, 1982). Critics have argued that both the concepts of 'innate' and 'instinctive' are slippery and vague (Bateson, 1996; Mameli and Bateson, 2011). Several meanings have been applied to these terms, including present at birth, genetically determined, unchanging during development, not influenced by learning, shared by all members of a species, and adapted over the course of evolution. Yet, these characteristics do not always align for a specific trait (Blumberg, 2017); for instance, some traits that are present at birth are influenced by learning, and some universally exhibited characteristics are not determined by genes. In humans, 'respect for private property' might be a characteristic that is found universally across all populations, but that does not mean that this behaviour is genetically determined, uninfluenced by learning, or unchanging during the lifespan. We suggest that the term 'human nature', along with the terms 'innate' and 'instinct', do more harm than good, as they constrain our thinking on the role of development in behaviour and cognition (Laland and Brown, 2018).

The idea that human nature can be contrasted with human culture is also inconsistent with the Cultural Evolution and Gene–Culture Coevolutionary perspectives, whereby neither genetic nor cultural inheritance is seen as independently reflecting the 'nature' of the human condition (Richerson, 2018). The concept is incompatible with the evidence that cultural niche construction has significantly impacted the evolution of our species (Laland and Brown, 2018). For instance, is cooking (a trait that is both universal and culturally learned) an aspect of human nature? And if human dentition, jaw musculature, and digestion have evolved in response to the cooking of our ancestors (Wrangham, 2009), then are these traits part of our nature, even though they are a consequence of our nurture? In sum, the validity and utility of a concept of a 'universal human nature' remains a point of contention within the field. If researchers commit to a definition of human nature as those aspects of the human mind that result from natural selection acting on genes, they might be missing perhaps the most important component of 'human nature', which is our propensity to engage in cultural transmission and underestimate the influence of culture on the evolutionary process. The dramatic impact of human culture on the evolutionary history of our species is perhaps what distinguishes us most clearly from other species.

Is there any rationale for Evolutionary Psychology maintaining a concept of human nature as a means of delineating the robust psychological tendencies that characterize our species? Evolutionary Psychologists have stood by the claim that developmental plasticity is simply the outcome of natural selection on genes, including where developmental outcomes are expressed in a probabilistic manner (e.g. Bjorkland, 2015; Nettle and Scott-Phillips, 2023; Tooby and Cosmides, 1990b, 1992). Evolved psychological mechanisms are argued to exhibit functional design that, in some instances, gives rise to derived functions such as the ability to read, but that the source of functional design always reflects natural selection acting on genes. In contrast, other researchers have argued that cultural evolution and developmental plasticity generate alternative sources of selection and design (e.g. Henrich, 2016; Heyes, 2018; Laland, 2017). The idea that human cognitive and behavioural outputs are adequately captured by a genetically determined 'reaction norm', whereby a range of specific behavioural options are predefined and delineated by genes, is considered too restrictive by researchers spanning many disciplines. We elaborate on this topic in the next section.

The extended evolutionary synthesis

Evolutionary theory is evolving. The change has been unfolding slowly, and the trajectory of travel has been resisted by some scientists, but the scientific field is changing, nonetheless. All areas of science undergo a continual process of development as new methods are devised, revised theories and hypotheses are proposed, and new empirical findings emerge. Such change can be dramatic, as the result of key discoveries or conceptual reformulations, but more frequently, scientific fields change at a slower pace, with ideas or theories taking some time to gain popularity with a new generation of researchers. Perhaps one such example is niche construction theory (described in Chapter 7). In brief, the term 'niche construction' was coined in 1988 by John Odling-Smee, who was first to argue that niche construction should be recognized as an evolutionary process. Richard Lewontin (1983) had previously written about organisms constructing niches and emphasized the importance of considering the evolutionary consequences of the actions of organisms on their environments, building on the earlier work of Conrad Waddington (1959) and others. The concept was brought to a wider audience over the subsequent decades, leading to the application of niche construction theory and related concepts in evolutionary biology, ecology, and the human sciences (e.g. Laland et al., 2000; Odling-Smee et al., 1996, 2003; Matthews et al., 2014;

Sultan, 2015). From slow beginnings, niche construction theory has provided novel insights into several topics in the human evolutionary sciences, including agriculture and domestication, human impacts on biodiversity, and human evolution (e.g. Boivin et al., 2016; Creanza et al., 2017; Wells and Stock, 2020; Zeder, 2017). The ongoing changes in evolutionary theory reflect slow accumulation of empirical evidence and a rethinking of the relationship between development and evolution.

Niche construction theory forms one part of the *extended evolutionary synthesis* (EES), which is a framework that focuses on the relationship between developmental and evolutionary processes. In addition to niche construction, the EES incorporates processes such as *developmental bias, plasticity-led evolution*, and *epigenetic inheritance*, thus presenting a broader version of evolutionary theory than was previously envisioned (Jablonka and Lamb, 2014; Lala et al., 2024; Laland et al., 2015; Müller, 2017; Pigliucci and Müller, 2010; West-Eberhard, 2003). Before describing some of these processes, let us briefly consider the standard model of evolutionary theory against which the EES might be compared. With the emergence of genetics as an academic field in the early twentieth century, Darwin's theory of natural selection could finally be tested based on a clearer understanding of the mechanisms of inheritance. This early framework, known as the *modern synthesis* (MS) (Huxley, 1942), became more elaborate and nuanced over time, as new methods were developed and empirical insights were gained. However, the standard evolutionary framework remained predominantly gene-focused, with the causal arrow of selection firmly pointing from the environment to the organism. Adaptations were defined as those traits that resulted from natural selection acting on genes, and inheritance was primarily restricted to gene transmission. The EES is an attempt to expand, rather than undermine, the MS by drawing greater attention to the bidirectional interactions between organisms and their environments (see Laland et al., 2015, for comparisons between the MS and EES). In the EES, the one-way arrow from genotypes to phenotypes is replaced by two-way arrows, where the activities of organism play a causal role in the evolutionary process.

Proponents of the EES have argued that neglecting the full breadth of causal processes leads to an impoverished view of how evolution works (Lala et al., 2024; Laland et al., 2014, 2015). For instance, many aspects of human evolution, including a more gracile skeleton, smaller teeth, reduced sexual dimorphism, and lower levels of aggression, can all be accounted for by natural selection acting on developmentally biased variation (Wilkins et al., 2014). The term 'developmental bias' refers to the idea that developmental systems are more likely to generate some characters, or trait combinations, than others.

'Plasticity-led evolution' relates to the idea that developmental plasticity can influence which phenotypes are exposed to selection and thus play a guiding role in genetic evolution. Some of the strongest evidence for plasticity-led evolution is found in our species: it is now widely accepted that our ancestors' domestication of plants and animals, and the associated changes in diet, led to genetic changes in human digestion (Henrich, 2016). The EES also stresses that extra-genetic forms of inheritance are important in evolution. 'Epigenetic inheritance' is one such type of extra-genetic inheritance that has been found to be widespread in nature, including in humans (Anastasiadi et al., 2021). It encompasses cases in which the experiences of parents during their lifespan (e.g. stress, diet, and exposure to toxins) have carry-over effects for their offspring (e.g. affecting body weight, stress priming, and disease incidence). Human culture is among the best-studied examples of extra-genetic inheritance, and, as discussed in Chapter 7, there is now evidence that it has elicited genetic change in our species.

Evolutionary biologists have differed in their views on whether the standard evolutionary framework requires a reconceptualization and, if so, the scale of this revision (Laland et al., 2014; Wray et al., 2014), which means that the EES remains a topic of some contention. However, few evolutionary biologists now dispute that developmental bias, plasticity-led evolution, extra-genetic inheritance, and niche construction are important in evolution, particularly in human evolution, and hence there are strong grounds for the human evolutionary sciences to take these topics seriously (e.g. Flynn et al., 2013; Packer and Cole, 2019; Stotz, 2014; Zeder, 2018). What are the specific implications of the EES for those studying the evolution of human behaviour and cognition? One point is that, from an EES perspective, the distinctions between Tinbergen's four questions about behaviour, which were mentioned earlier in this chapter (also in Chapter 2), become more blurred (Bateson and Laland, 2013); for example, asking questions about the *development* of a trait becomes relevant to providing answers about the *evolutionary history* of that trait. Similarly, from an EES perspective, the distinction between *proximate* causes of behaviour (e.g. genes, hormones, and developmental processes) and *ultimate* causes of behaviour (e.g. natural selection, genetic drift) is not so clean, since processes traditionally regarded as proximate mechanisms (e.g. learning, culture) can be involved in ultimate causation (Laland et al., 2011). The gene-centric view of evolution, which has formed a core part of standard evolutionary theory for many decades (Ågren, 2021), underestimates the causal influence of developmental processes, including learned behaviours, which is likely to be particularly relevant for understanding the evolution of our own species. Any suggestion that human behaviour and cognition is solely the outcome of natural selection acting on genes (e.g. Baumard et al., 2024; Nettle and Scott-Phillips, 2023) is disputed by the EES (Laland et al., 2015). Even the 'reaction norm' metaphor, which suggests that phenotypes are constrained to a particular set of genetically controlled outputs, is inconsistent with an EES framework, which emphasizes that phenotypic novelty is produced by developmental processes, triggering plasticity-led evolution and facilitating extra-genetic inheritance.

Conclusion

Our goal has been to provide readers with an insight into the human evolutionary behavioural sciences and accurately present the contemporary subfields while also taking a critical view of each approach. We hope to have given a taste of the variety of methods that are currently applied, thereby providing a 'guide for the bewildered'. Given the diverse backgrounds and interests of the practitioners, it is hardly surprising that distinct perspectives on human behaviour have emerged, largely reflecting the methodological and conceptual habits of the parent disciplines. Inevitably, the different approaches are sometimes seen as providing competing views of human behaviour. However, when these alternatives are examined more closely it becomes clear that there is little that is genuinely incompatible about their explanations or methodologies. While there are some theoretical differences, we predict that these differences will eventually be settled by empirical research. In the meantime, as the case study on infanticide demonstrated, there is a complementarity of information generated by the different methods. Individual researchers are free to draw from different styles to synthesize their own integrative and pluralistic evolutionary analyses. Points of contention remain within this field, for instance, concerning the role of culture in the evolutionary process, the utility of 'human nature' concepts, and the need for an extended evolutionary synthesis. This scientific field, like all science, will continue to evolve. Our historical perspective has emphasized the damage done by the misapplication of evolutionary theory. Newspapers, magazines, and online fora continue to propagate sensationalist, and often unsupported, claims about the evolution of human behaviour, which we hope the reader is now better placed to evaluate for themselves.

We should not be surprised if differences of opinion persist within science. As we have seen, arguments have broken out between the different subfields, and occasionally the exchanges have become quite heated. The past decades have witnessed repeated pleas in the human behaviour and evolution literature for the 'rival factions' to settle their differences, to dwell on their common ground, and to bond in the face of the considerable hostility that is seen as emanating from the massed ranks of the social sciences. Yet the mere assertion that 'we're all in this together' is unlikely to generate a true integration of perspectives when particular camps regard the alternatives as out of touch with relevant literatures or methodologically weaker than themselves. We do not agree with the view that it is divisive to dwell on any differences of methodology or conviction within the field, nor that it is damaging for the discipline to be exacting. On the contrary, to the extent that human sociobiology and its descendants have consciously or inadvertently discouraged a self-critical ethos among their practitioners, we regard this resistance as a barrier to integration. Why should researchers want the respectable, disciplined work of their subdiscipline to be associated with the superficial analyses of other researchers? Inevitably, they will fear that inflammatory declarations, careless popularizations, and adaptationist storytelling will produce a backlash against all evolutionary approaches in the social sciences. Given the valuable research on human behaviour and evolution carried out by all the sub-fields, this outcome would not only be a great shame but an unnecessary tragedy. The field of evolution and human behaviour is no longer a vulnerable sapling but has developed into a vibrant and vigorously growing tree with roots sufficiently well established for it to be able to stand up to, and indeed benefit from, healthy pruning.

A delicate balance must be struck here. While there is a need for genuine pluralism of methodology, it does not follow that all analyses based loosely on evolutionary conjecture are salutary. High standards of research, rather than unconditional positive regard to fellow enthusiasts, are the best defence against external disapprobation. What is needed is a pluralistic yet rigorous, fertile yet self-critical, scientific discipline that at the same time champions bona fide evolutionary methods and inferences but clamps down hard on undisciplined storytelling and potentially damaging or abusive evolutionary reasoning. A genuine marriage of the biological and social sciences will only emerge when the ratio of sense to nonsense is improved.

References

Acerbi, A. and Mesoudi, A. (2015). If we are all cultural Darwinian's what's the fuss about? Clarifying recent disagreements in the field of cultural evolution. *Biology and Philosophy* 30: 481–503.

Ågren, J. A. (2021). *The Gene's-Eye View of Evolution*. Oxford, UK: Oxford University Press.

Alexander, R. D. (1974). The evolution of social behavior. *Annual Review of Ecology and Systematics* 5: 325–383.

Alexander, R. D. (1979). *Darwinism and Human Affairs*. London, UK: Pitman.

Alger, I., Hooper, P. L., Cox, D., Stieglitz, J. and Kaplan, H. S. (2020). Paternal provisioning results from ecological changes. *Proceedings of the National Academy of Sciences* 117: 10746–10754.

Allen, E., Beckwith, B., Beckwith, J., Chorover, S., Culver, D., Duncan, M., Gould, S., Hubbard, R., et al. (1975). Against 'Sociobiology' [Letter]. *New York Review of Books* 182: 184–186.

Alvergne, A., Faurie, C. and Raymond, M. (2009). Father–offspring resemblance predicts paternal investment in humans. *Animal Behaviour* 78: 61–69.

Alvergne, A., Faurie, C. and Raymond, M. (2010). Are parents' perceptions of offspring facial resemblance consistent with actual resemblance? Effects on parental investment. *Evolution and Human Behavior* 31: 7–15.

Anastasiadi, D., Venney, C. J., Bernatchez, L. and Wellenreuther, M. (2021). Epigenetic inheritance and reproductive mode in plants and animals. *Trends in Ecology and Evolution* 36: 1124–1140.

Anderson, M. L. (2010). Neural reuse: a fundamental organizational principle of the brain. *Behavioral and Brain Sciences* 33: 245–266.

Anderson, M. L. and Finlay, B. L. (2014). Allocating structure to function: the strong links between neuroplasticity and natural selection. *Frontiers in Human Neuroscience* 7: 918.

Aoki, K. (1986). A stochastic model of gene–culture coevolution suggested by the 'culture historical hypothesis' for the evolution of adult lactose absorption in humans. *Proceedings of the National Academy of Sciences* 83: 2929–2933.

Apicella, C., Norenzayan, A. and Henrich, J. (2020). Beyond WEIRD: a review of the last decade and a look ahead to the global laboratory of the future. *Evolution and Human Behavior* 41: 319–329.

Aplin, L. (2019). Culture and cultural evolution in birds: a review of the evidence. *Animal Behaviour* 147: 179–187.

Archer, J. (2013). Can evolutionary principles explain patterns of family violence? *Psychological Bulletin* 139: 403–440.

Ardrey, R. (1966). *The Territorial Imperative*. London, UK: Collins.

Arendt, M., Cairns, K. M., Ballard, J. W. O., Savolainen, P. and Axelsson, E. (2016). Diet adaptation in dog reflects spread of prehistoric agriculture. *Heredity* 117: 301–306.

Atkinson, A. P. and Wheeler, M. (2004). The grain of domains: the evolutionary psychological case against domain–general cognition. *Mind and Language* 19: 147–176.

Atkinson, E. G., Audesse, A. J., Palacios, J. A., Bobo, D. M., Webb, A. E., Ramachandran, S. and Henn, B. M. (2018). No evidence for recent selection at FOXP2 among diverse human populations. *Cell* 174: 1424–1435.

Atkinson, Q. D., Meade, A. Venditti, C., Greenhill, S. J. and Pagel, M. (2008). Language evolves in punctuational bursts. *Science* **319**: 588.
Aubert-Teillaud, B. and Vaidis, D. C. (2024). Conspiracy theories explained by a cheating detection mechanism. *Social and Personality Psychology Compass* **18**: e12876.
Aunger, R. (Ed.) (2000a). *Darwinizing Culture: The Status of Memetics as a Science*. Oxford, UK: Oxford University Press.
Aunger, R. (2000b). The life history of culture learning in a face-to-face society. *Ethos* **28**: 1–38.
Bagemihl, B. (1999). *Biological Exuberance: Animal Homosexuality and Natural Diversity*. London, UK: Profile Books.
Bailey, N. W. and Zuk, M. (2009). Same-sex sexual behavior and evolution. *Trends in Ecology and Evolution* **24**: 439–446.
Bandini, E. and Harrison, R. A. (2020). Innovation in chimpanzees. *Biological Reviews* **95**: 1167–1197.
Barkow, J. H., Cosmides, L. and Tooby, J. (Eds.) (1992). *The Adapted Mind: Evolutionary Psychology and the Generation of Culture*. Oxford, UK: Oxford University Press.
Barrett, H. C. (2007). Modules in the flesh. In S. Gangestad and J. Simpson (Eds.) *The Evolution of Mind* (pp. 161–168). New York, NY: Guilford Press.
Barrett, H. C. (2015). *The Shape of Thought: How Mental Adaptations Evolve*. Oxford, UK: Oxford University Press.
Barrett, H. C. (2020). The search for human cognitive specializations. In J. H. Kaas (Ed.) *Evolutionary Neuroscience* (2nd edition) (pp. 917–929). Cambridge, MA: Academic Press.
Barrett, H. C. (2024). Evolutionary psychology. In J. Koster, B. Scelza and M. K. Shenk (Eds.) *Human Behavioral Ecology* (pp. 380–401). Cambridge, UK: Cambridge University Press.
Barrett, H. C. and Armstrong, J. (2023). Climate change adaptation and the back of the invisible hand. *Philosophical Transactions of the Royal Society B* **378**: 20220406.
Barrett, L., Pollet, T. V. and Stulp, G. (2014). From computers to cultivation: reconceptualizing evolutionary psychology. *Frontiers in Psychology* **5**: 867.
Basalla, G. (1988). *The Evolution of Technology*. Cambridge, UK: Cambridge University Press.
Bateman, A. J. (1948). Intra-sexual selection in *Drosophila*. *Heredity* **2**: 349–368.
Bateson, P. (1994). The dynamics of parent–offspring relationships in mammals. *Trends in Ecology and Evolution* **9**: 399–402.
Bateson, P. (1996). Design for a life. In D. Magnusson (Ed.) *The Lifespan Development of Individuals* (pp. 1–20). Cambridge, UK: Cambridge University Press.
Bateson, P. (2017). *Behaviour, Development and Evolution*. Cambridge, UK: Open Book Publishers.
Bateson, P. and Laland, K. N. (2013). Tinbergen's four questions: an appreciation and an update. *Trends in Ecology and Evolution* **28**: 712–718.
Bateson, P. and Martin, P. (1999). *Design for a Life: How Behaviour Develops*. London, UK: Jonathan Cape.
Baumard, N., André, J., Nettle, D., Fitouchi, L. and Scott-Philipps, T. (2024). The gene's-eye view of culture: vehicles, not replicators. In M. L. Fisher (Ed.), *APA Handbook of Evolutionary Psychology*. Washington, DC: APA Press. https://doi.org/10.32942/X2MS51
Bell, A. V., Richerson, P. J. and McElreath, R. (2009). Culture rather than genes provides greater scope for the evolution of large-scale human prosociality. *Proceedings of the National Academy of Sciences* **106**: 17671–17674.
Bell, R. and Buchner, A. (2012). How adaptive is memory for cheaters? *Current Directions in Psychological Science* **21**: 403–408.
Benedict, R. (1924). *Patterns of Culture*. Boston, MA: Houghton Mifflin (reprinted 1989, Boston, MA: Houghton Mifflin).

Benn Torres, J. (2020). Anthropological perspectives on genomic data, genetic ancestry, and race. *Yearbook of Physical Anthropology* **171** (Supplement 70): 74–86.
Ben-Oren, Y., Kolodny, O. and Creanza, N. (2023). Cultural specialization as a double-edged sword: division into specialized guilds might promote cultural complexity at the cost of higher susceptibility to cultural loss. *Philosophical Transactions of the Royal Society B* **378**: 20210418.
Bergström, A., Stringer, C., Hajdinjak, M., Scerri, E. M. L. and Skoglund, P. (2021). Origins of modern human ancestry. *Nature* **590**: 229–237.
Bettinger, R. L. and Eerkens, J. (1999). Point typologies, cultural transmission, and the spread of bow-and-arrow technology in the prehistoric Great Basin. *American Antiquity* **64**: 231–242.
Bickel, B., Witzlack-Makarevich, A., Choudhary, K. K., Schlesewsky, M. and Bornkessel-Schlesewsky, I. (2015). The neurophysiology of language processing shapes the evolution of grammar: evidence from case marking. *PLoS One* **10**: e0132819.
Birch, J. and Heyes, C. (2021). The cultural evolution of cultural evolution. *Philosophical Transactions of the Royal Society B* **376**: 20200051.
Biro, D., Haslam, M. and Rutz, C. (2013). Tool use as adaptation. *Philosophical Transactions of the Royal Society B* **368**: 20120408.
Bjorklund, D. F. (2015). Developing adaptations. *Developmental Review* **38**: 13–35.
Bjorklund, D. F. (2016). Prepared is not preformed: commentary on Witherington and Lickliter. *Human Development* **59**: 235–241.
Bjorklund, D. F. and Pellegrini, A. D. (2000). Child development and evolutionary psychology. *Child Development* **71**: 1687–1708.
Bjorklund, D. F., Hernández Blasi, C. and Ellis, B. J. (2015). Evolutionary developmental psychology. In D. M. Buss (Ed.) *The Handbook of Evolutionary Psychology* (2nd edition) (pp. 904–925). Hoboken, NJ: Wiley.
Björklund, M. (2019). Lamarck, the father of evolutionary ecology? *Trends in Ecology and Evolution* **34**: 874–875.
Blackmore, S. (1999). *The Meme Machine*. Oxford, UK: Oxford University Press.
Blackwell, A. B. (1875). *The Sexes Throughout Nature*. New York, NY: G. P. Putnam.
Bloch, M. (2000). A well-disposed social anthropologist's problems with memes. In R. Aunger (Ed.) *Darwinizing Culture: The Status of Memetics as a Science* (pp. 189–203). Oxford, UK: Oxford University Press.
Blumberg, M. S. (2017). Development evolving: the origins and meanings of instinct. *WIREs Cognitive Science* **8**: e1371.
Blurton Jones, N. (1986). Bushman birth spacing: a test for optimal interbirth interval. *Ethology and Sociobiology* **7**: 91–105.
Blurton Jones, N. (1997). Too good to be true? Is there really a trade-off between number and care of offspring in human reproduction? In L. Betzig (Ed.) *Human Nature: A Critical Reader* (pp. 83–86). Oxford, UK: Oxford University Press.
Blurton Jones, N. and Sibly, R. M. (1978). Testing adaptive-ness of culturally determined behaviour: do bushman women maximise their reproductive success by spacing births widely and foraging seldom? In N. Blurton Jones and V. Reynolds (Eds.) *Human Behaviour and Adaptation* (pp. 135–158). London, UK: Francis Taylor.
Boakes, R. (1984). *From Darwin to Behaviourism: Psychology and the Minds of Animals*. Cambridge, UK: Cambridge University Press.
Bobrow, D. and Bailey, J. M. (2001). Is male homosexuality maintained via kin selection? *Evolution and Human Behavior* **22**: 361–368.
Boivin, N. L., Zeder, M. A., Fuller, D. Q., Crowther, A., Larson, G., Erlandson, J. M., Denham, T. and Petraglia, M. D. (2016). Ecological consequences of human niche

construction: examining long-term anthropogenic shaping of global species distributions. *Proceedings of the National Academy of Sciences* 113: 6388–6396.

Bolhuis, J. J. and MacPhail, E. M. (2001). A critique of the neuroecology of learning and memory. *Trends in Cognitive Sciences* 5: 426–433.

Bolhuis, J. J., Brown, G. R., Richardson, R. C. and Laland, K. N. (2011). Darwin in mind: new opportunities for evolutionary psychology. *PLoS Biology* 9: e1001109.

Bonduriansky, R. and Day, T. (2018). *Extended Heredity: A New Understanding of Inheritance and Evolution*. Princeton, NJ: Princeton University Press.

Bonnet, T., Morrissey, M. B., de Villemeureuil, P., Alberts, S. C., Arcese, P., Bailey, L. D., Boutin, S., Brekke, P., et al. (2022). Genetic variance in fitness indicates rapid contemporary adaptive evolution in wild animals. *Science* 376: 1012–1016.

Borgerhoff Mulder, M. (1990). Kipsigis women's preference for wealthy men: evidence for female choice in mammals? *Behavioral Ecology and Sociobiology* 27: 255–264.

Borgerhoff Mulder, M. (1991). Human behavioural ecology. In J. R. Krebs and N. B. Davies (Eds.) *Behavioural Ecology: An Evolutionary Approach* (3rd edition) (pp. 69–98). Oxford, UK: Blackwell Scientific Publications.

Borgerhoff Mulder, M. (2000). Optimizing offspring: the quantity–quality tradeoff in agropastoral Kipsigis. *Evolution and Human Behavior* 21: 391–410.

Borgerhoff Mulder, M. (2013). Human behavioral ecology—necessary but not sufficient for the evolutionary analysis of human behaviour. *Behavioral Ecology* 24: 1042–1043.

Borgerhoff Mulder, M. (2024). Bateman's principles and the study of evolutionary demography. In O. Burger, R. Lee and R. Sear (Eds.) *Human Evolutionary Demography* (pp. 551–574). Cambridge, UK: Open Book Publishers.

Borgerhoff Mulder, M., Bowles, S., Hertz, T., Bell, A., Beise, J., Clark, G., Fazzio, I., Gurven, M., et al. (2009). Intergenerational wealth transmission and the dynamics of inequality in small-scale societies. *Science* 326: 682–688.

Borgerhoff Mulder, M. and Schacht, R. (2012). Human behavioural ecology. In eLS. Chichester, UK: Wiley. https://doi.org/10.1002/9780470015902.a0003671.pub2

Bowden, R., MacFie, T. S., Myers, S., Hellenthal, G., Nerrienet, E., Bontrop, R. E., Freeman, C., Donnelly, P., et al. (2012). Genomic tools for evolution and conservation in the chimpanzee: *Pan troglodytes ellioti* is a genetically distinct population. *PLoS Genetics* 8: e1002504.

Bowlby, J. (1969). *Attachment and Loss: Volume 1: Attachment*. London, UK: Hogarth Press.

Bowles, S. (2009). Did warfare among ancestral hunter-gatherers affect the evolution of human social behaviors? *Science* 324: 1293–1298.

Bowles, S. and Gintis, H. (2004). The evolution of strong reciprocity: cooperation in heterogeneous populations *Theoretical Population Biology* 65: 17–28.

Boyd, R., Gintis, H. and Bowles, S. (2010). Coordinated punishment of defectors sustains cooperation and can proliferate when rare. *Science* 328: 617–620.

Boyd, R., Gintis, H., Bowles, S. and Richerson, P. J. (2003). The evolution of altruistic punishment. *Proceedings of the National Academy of Sciences* 100: 3531–3535.

Boyd, R. and Richerson, P. (1983). The cultural transmission of acquired variation: effects on genetic fitness. *Journal of Theoretical Biology* 100: 567–596.

Boyd, R. and Richerson, P. J. (1985). *Culture and the Evolutionary Process*. Chicago, IL: Chicago University Press.

Boyd, R. and Richerson, P. J. (1995). Why does culture increase human adaptability? *Ethology and Sociobiology* 16: 125–143.

Boyer, P. (1994). *Naturalness of Religious Ideas*. Berkeley, CA: University of California Press.

Brase, G. L., Cosmides, L. and Tooby, J. (1998). Individuation, counting, and statistical inference: the role of frequency and whole-object representations in judgment under uncertainty. *Journal of Experimental Psychology: General* 127: 3–21.

Bribiescas, R. G. (2016). *How Men Age: What Evolution Reveals About Male Health and Mortality*. Princeton, NJ: Princeton University Press.

Brigham, C. C. (1923). *A Study of American Intelligence*. Princeton, NJ: Princeton University Press.

Brochhagen, T., Boleda, G., Gualdoni, E. and Xu, Y. (2023). From language development to language evolution: a unified view of human lexical creativity. *Science* 381: 431–436.

Brodie, R. (1996). *Virus of the Mind: The New Science of the Meme*. Seattle, WA: Integral Press.

Brooke, J. H. (2001). The Wilberforce–Huxley debate: why did it happen? *Science and Christian Belief* 13: 127–141.

Brown, D. E. (1991). *Human Universals*. New York, NY: McGraw-Hill.

Brown, G. R. (2001). Sex-biased investment in nonhuman primates: can Trivers and Willard's theory be tested? *Animal Behaviour* 61: 683–694.

Brown, G. R. (2013). Why mechanisms shouldn't be ignored: commentary on Nettle et al.'s 'Human behavioural ecology: current research and future prospects'. *Behavioral Ecology* 24: 1041–1042.

Brown, G. R., Laland, K. N. and Borgerhoff-Mulder, M. (2009). Bateman's principles and human sex roles. *Trends in Ecology and Evolution* 24: 297–304.

Brown, G. R. and Silk, J. B. (2002). Reconsidering the null hypothesis: is maternal rank associated with birth sex ratios in primate groups? *Proceedings of the National Academy of Sciences* 99: 11252–11255.

Buller, D. J. (2005). Evolutionary psychology: the emperor's new paradigm. *Trends Cognitive Sciences* 9: 277–283.

Bulmer, M. (2003). *Francis Galton: Pioneer of Heredity and Biometry*. Baltimore, MD: John Hopkins University Press.

Burkart, J. M., Schubiger, M. N. and van Schaik, C. P. (2017). The evolution of general intelligence. *Behavioral and Brain Sciences* 40: e195.

Burkhardt, R. W. (2005). *Patterns of Behavior: Konrad Lorenz, Niko Tinbergen, and the Founding of Ethology*. Chicago, IL: University of Chicago Press.

Buskell, A., Enquist, M. and Jansson, F. (2019). A systems approach to cultural evolution. *Palgrave Communications* 5: 131.

Buss, D. M. (1994). *The Evolution of Desire: Strategies of Human Mating*. New York, NY: HarperCollins.

Buss, D. M. (1995). Evolutionary psychology: a new paradigm for psychological science. *Psychological Inquiry* 6: 1–30.

Buss, D. M. (1999). *Evolutionary Psychology: The New Science of the Mind*. London, UK: Allyn and Bacon.

Buss, D. M. (Ed.) (2005). *The Handbook of Evolutionary Psychology* (1st edition). Hoboken, NJ: Wiley.

Buss, D. M. (2008). *Evolutionary Psychology: The New Science of the Mind* (3rd edition). New York, NY: Pearson.

Buss, D. M. (Ed.) (2015). *The Handbook of Evolutionary Psychology* (2nd edition; two volumes). Hoboken, NJ: Wiley.

Buss, D. M. (2018). Sexual and emotional infidelity: evolved gender differences in jealousy prove robust and replicable. *Perspectives on Psychological Science* 13: 155–160.

Buss, D. M. (2019). *Evolutionary Psychology: The New Science of the Mind* (6th edition). New York, NY: Routledge.

Buss, D. M., Abbott, M., Angleitner, A., Asherian, A., Biaggio, A., Blanco-Villasenor, A., Bruchon-Schweitzer, M., Ch'u, H., et al. (1990). International preferences in selecting mates: a study of 37 cultures. *Journal of Cross-Cultural Psychology* 21: 5–47.

Caldwell, C. A., Atkinson, M. and Renner, E. (2016). Experimental approaches to studying cumulative cultural evolution. *Current Directions in Psychological Science* 24: 191–195.

Caldwell, C. A. and Millen, A. E. (2008). Experimental models for testing hypotheses about cumulative cultural evolution. *Evolution and Human Behavior* **29**: 165–171.
Caldwell, C. A. and Millen, A. E. (2009). Social learning mechanisms and cumulative cultural evolution: is imitation necessary? *Psychological Science* **12**: 1478–1453.
Call, J., Burghardt, G., Pepperberg, I., Snowdon, C. and Zentall, T. R. (Eds.) (2017). *APA Handbook of Comparative Psychology* (Volumes 1 and 2). Washington, DC: American Psychological Association.
Caneiro, R. L. (2003). *Evolutionism in Cultural Anthropology*. Boulder, CO: Westview Press.
Cäsar, C. and Zuberbühler, K. (2012). Referential alarm calling behaviour in New World primates. *Current Zoology* **58**: 680–697.
Cavalli-Sforza, L. L. and Feldman, M. W. (1981). *Cultural Transmission and Evolution: A Quantitative Approach*. Princeton, NJ: Princeton University Press.
Cavalli-Sforza, L. L., Feldman, M. W., Chen, K. H. and Dornbusch, S. M. (1982). Theory and observation in cultural transmission. *Science* **218**: 19–27.
Cavalli-Sforza, L. L., Minch, E. and Mountain, J. L. (1992). Coevolution of genes and languages revisited. *Proceedings of the National Academy of Science* **89**: 6020–6024.
Cavalli-Sforza, L. L. and Wang, W. S.-Y. (1986). Spatial distance and lexical replacement. *Language* **62**: 38–55.
Chagnon, N. A. and Irons, W. (Eds.) (1979). *Evolutionary Biology and Human Social Behavior: An Anthropological Perspective*. North Scituate, MA: Duxbury Press.
Chapman, S. N., Lahdenperä, M. Pettay, J. E., Lynch, R. F. and Lummaa, V. (2021). Offspring fertility and grandchild survival enhanced by maternal grandmothers in a pre-industrial human society. *Scientific Reports* **11**: 3652.
Charlesworth, B. and Charlesworth, D. (2003). *Evolution: A Very Short Introduction*. Oxford, UK: Oxford University Press.
Chomsky, N. (1965). *Aspects of the Theory of Syntax*. Cambridge, MA: MIT Press.
Claidière, N., Scott-Phillips, T. C. and Sperber, D. (2014). How Darwinian in cultural evolution? *Philosophical Transactions of the Royal Society B* **369**: 20130368.
Claidière, N. and Sperber, D. (2007). The role of attraction in cultural evolution. *Journal of Cognition and Culture* **7**: 89–111.
Clark, A. D., Deffner, D., Laland, K., Odling-Smee, J. and Endler, J. (2020). Niche construction affects the variability and strength of natural selection. *American Naturalist* **195**: 16–30.
Clarke, G. M. (1995). Relationships between developmental stability and fitness: applications for conservation biology. *Conservation Biology* **9**: 18–24.
Clarke, J. I. (2000). *The Human Dichotomy : The Changing Numbers of Males and Females*. Amsterdam, The Netherlands: Pergamon Press.
Clutton-Brock, T. (2009). Cooperation between non-kin in animal societies. *Nature* **462**: 51–57.
Clutton-Brock, T. H. and Parker, G. A. (1995). Sexual coercion in animal societies. *Animal Behaviour* **49**: 1345–1365.
Codding, B. F., Bliege Bird, R. and Bird, D. W. (2011). Provisioning offspring and others: risk-energy trade-offs and gender differences in hunter-gatherer foraging strategies. *Proceedings of the Royal Society B* **278**: 2502–2509.
Colleran, H. (2016). The cultural evolution of fertility decline. *Philosophical Transactions of the Royal Society B* **371**: 20150152.
Conroy-Beam, D., Buss, D. M., Asao, K., Sorokowska, A., Sorokowski, P., Aavil, T., Akello, G., Alhabahda, M. M., et al. (2019). Contrasting computational models of mate preference integration across 45 countries. *Scientific Reports* **9**: 16885.
Cook, M. and Mineka, S. (1989). Observational conditioning of fear to fear-relevant versus fear-irrelevant stimuli in rhesus monkeys. *Journal of Abnormal Psychology* **98**: 448–459.

Corado, D. T., Sun, R., Keltner, D., Kamble, S., Huddar, N. and McNeil, G. (2018). Universals and cultural variations in 22 emotional expressions across five cultures. *Emotion* 18: 75–93.

Cosmides, L. (1989). The logic of social exchange: has natural selection shaped how humans reason? Studies with the Wason selection task. *Cognition* 31: 187–276.

Cosmides, L., Barrett, H. C. and Tooby, J. (2010). Adaptive specializations, social exchange, and the evolution of human intelligence. *Proceedings of the National Academy of Sciences* 107: 9007–9014.

Cosmides, L. and Tooby, J. (1987). From evolution to behavior: evolutionary psychology as the missing link. In J. Dupré (Ed.) *The Latest on the Best: Essays on Evolution and Optimality* (pp. 276–306). Cambridge, MA: MIT Press.

Cosmides, L. and Tooby, J. (1992). Cognitive adaptations for social exchange. In J. H. Barkow, L. Cosmides and J. Tooby (Eds.) *The Adapted Mind: Evolutionary Psychology and the Generation of Culture* (pp. 163–228). Oxford, UK: Oxford University Press.

Cosmides, L. and Tooby, J. (2000). Evolutionary psychology and the emotions. In M. Lewis and J. M. Haviland-Jones (Eds.) *Handbook of Emotions* (2nd edition) (pp. 91–115). New York, NY: Guilford Press.

Costall, A. (1993). How Lloyd Morgan's canon backfired. *Journal of the History of the Behavioral Sciences* 29: 113–122.

Creanza, N. and Feldman, M. W. (2016). Worldwide genetic and cultural change in human evolution. *Current Opinion in Genetics and Development* 41: 85–92.

Creanza, N., Kolodny, O. and Feldman, M. W. (2017). Cultural evolutionary theory: how culture evolves and why it matters. *Proceedings of the National Academy of Sciences* 114: 7782–7789.

Cronk, L. (1991). Human behavioral ecology. *Annual Review of Anthropology* 20: 25–53.

Cronk, L., Chagnon, N. and Irons, W. (Eds.) (2000). *Adaptation and Human Behavior: An Anthropological Perspective*. New York, NY: Aldine de Gruyter.

Crook, J. and Crook, S. J. (1988). Tibetan polyandry: problems of adaptation and fitness. In L. Betzig, M. Borgerhoff Mulder and P. Turke (Eds.) *Human Reproductive Behaviour: A Darwinian Perspective* (pp. 97–114). Cambridge, UK: Cambridge University Press.

Crook, J. and Gartlan, J. S. (1966). Evolution of primate societies. *Nature* 210: 1200–1203.

Cross, C. P. and Campbell, A. (2011). Women's aggression. *Aggression and Violent Behavior* 16: 390–398.

Cross, C. P., Copping, L. T. and Campbell, A. (2011). Sex differences in impulsivity: a meta-analysis. *Psychological Bulletin* 137: 97–130.

Currie, T. E., Campenni, M., Flitton, A., Njagi, T., Ontiri, E., Perret, C. and Walker, L. (2021). The cultural evolution and ecology of institutions. *Philosophical Transactions of the Royal Society B* 376: 20200047.

Currie, T. E., Greenhill, S. J., Gray, R. D., Hasegawa, T. and Mace, R. (2010a). Rise and fall of political complexity in island South-East Asia and the Pacific. *Nature* 467: 801–804.

Currie, T. E., Greenhill, S. J. and Mace, R. (2010b). Is horizontal transmission really a problem for phylogenetic comparative methods? A simulation study using continuous cultural traits. *Philosophical Transactions of the Royal Society B* 365: 3903–3912.

Curtis, V., Aunger, R. and Rabie, T. (2004). Evidence that disgust evolved to protect for risk of disease. *Proceedings of the Royal Society London, Series B* 271: S131–S133.

Curtis, V., de Barra, M. and Aunger, R. (2011). Disgust as an adaptive system for disease avoidance behaviour. *Philosophical Transactions of the Royal Society B* 366: 389–401.

Curtis, V. and Biran, A. (2001). Dirt, disgust, and disease: is hygiene in our genes? *Perspectives in Biology and Medicine* 44: 17–31.

Daly, M. and Wilson, M. I. (1982). Whom are newborn babies said to resemble? *Ethology and Sociobiology* 3: 69–78.

Daly, M. and Wilson, M. (1983). *Sex, Evolution and Behavior* (2nd edition). Belmont, CA: Wadsworth.

Daly, M. and Wilson, M. (1988). *Homicide*. New York, NY: Aldine.

Daly, M. and Wilson, M. I. (1999). Human evolutionary psychology and animal behaviour. *Animal Behaviour* 57: 509–519.

Daly, M. and Wilson, M. (2008). Is the 'Cinderella effect' controversial? A case study of evolution-minded research and critiques thereof. In C. Crawford and D. Krebs (Eds.) *Foundations of Evolutionary Psychology* (pp. 383–400). New York, NY: Taylor and Francis.

Danchin, É., Giraldeau, L.-A. and Cézilly, F. (Eds.) (2008). *Behavioural Ecology*. Oxford, UK: Oxford University Press.

Darwin, C. (1859). *On The Origin of Species by Means of Natural Selection, or the Preservation of Favoured Races in the Struggle for Life*. London, UK: John Murray (1st edition reprinted 1968, London, UK: Penguin Books).

Darwin, C. (1871). *The Descent of Man, and Selection in Relation to Sex*. London, UK: John Murray (1st edition reprinted 1981, Princeton NJ: Princeton University Press).

Darwin, C. (1872). *The Expression of the Emotions in Man and Animals*. London, UK: John Murray (3rd edition reprinted 1998, London, UK: Harper-Collins).

Davies, N. B., Krebs, J. R. and West, S. A. (2012). *An Introduction to Behavioural Ecology*. Chichester, UK: Wiley-Blackwell.

Dawkins, R. (1976). *The Selfish Gene*. Oxford, UK: Oxford University Press.

Dawkins, R. (1981). Selfish genes in race or politics. *Nature* 289: 528.

Dawkins, R. (1982). *The Extended Phenotype*. Oxford, UK: Oxford University Press.

DeBruine, L. M., Hahn, A. C. and Jones, B. C. (2016). Perceiving infant faces. *Current Opinion in Psychology* 7: 87–91.

DeCasien, A. R., Barton, R. A. and Higham, J. P. (2022). Understanding the human brain: insights from comparative biology. *Trends in Cognitive Sciences* 26: 432–445.

Deffner, D., Kleinow, V. and McElreath, R. (2020). Dynamic social learning in temporally and spatially variable environments. *Royal Society Open Science* 7: 200734.

Deffner, D. and McElreath, R. (2022). When does selection favor learning from the old? Social learning in age-structured populations. *PLoS One* 17: e0267204.

Deffner, D., Rohrer, J. M. and McElreath, R. (2022). A causal framework for cross-cultural generalizability. *Advances in Methods and Practices in Psychological Science* 5: 1–18.

Del Giudice, M., Ellis, B. J. and Shirtcliff, E. A. (2011). The adaptive calibration model of stress responsivity. *Neuroscience and Biobehavioral Reviews* 35: 1562–1592.

Del Giudice, M., Gangestad, S. W. and Kaplan, H. S. (2015). Life history theory and evolutionary psychology. In D. M. Buss (Ed.), *The Handbook of Evolutionary Psychology (Volume 1: Foundations (2nd ed.)* (pp. 88–114). New York, NY: Wiley.

Dennett, D. (1991). *Consciousness Explained*. London, UK: Penguin Books.

Dennett, D. (1995). *Darwin's Dangerous Idea: Evolution and the Meanings of Life*. London, UK: Penguin Books.

Denton, K. K., Ram, Y. and Feldman, M. W. (2023). Conditions that favour cumulative cultural evolution. *Philosophical Transactions of the Royal Society B* 378: 20210400.

Derex, M., Bonnefon, J., Boyd, R. and Mesoudi, A. (2019). Causal understanding is not necessary for the improvement of culturally evolving technology. *Nature Human Behaviour* 3: 446–452.

DeSilva, J. M., Traniello, J. F. A., Claxton, A. G. and Fannin, L. D. (2021). When and why did human brain decrease in size? A new change-point analysis and insights from brain evolution in ants. *Frontiers in Ecology and Evolution* 9: 742639.

Desmond, A. (1997). *Huxley: Evolution's High Priest*. London, UK: Michael Joseph.

Desmond, A. and Moore, J. (2009). *Darwin's Sacred Cause: Race, Slavery and the Quest for Human Origins*. London, UK: Penguin.

Dewsbury, D. (1995). Americans in Europe: the role of travel in the spread of European ethology after World War II. *Animal Behaviour* 49: 1649–1663.

Diamond, J. (1991). *The Rise and Fall of the Third Chimpanzee*. London, UK: Vintage.

Dickemann, M. (1979). Female infanticide, reproductive strategies, and social stratification: a preliminary model. In N. A. Chagnon and W. Irons (Eds.) *Evolutionary Biology and Human Social Behavior: An Anthropological Perspective* (pp. 321–367). North Scituate, MA: Duxbury Press.

Dixson, A. F. (2012). *Primate Sexuality: Comparative Studies of the Prosimians, Monkeys, Apes, and Humans* (2nd edition). Oxford, UK: Oxford University Press.

Dobzhansky, T. (1937). *Genetics and the Origin of Species*. New York, NY: Columbia University Press.

Donald, D. and Munro, J. (2009). *Endless Forms: Charles Darwin, Natural Selection and the Visual Arts*. New Haven, CT: Yale University Press.

Drummond, H. (1894). *The Ascent of Man*. London, UK: Hodder and Stoughton.

van Duijn, M. (2017). Phylogenetic origins of biological cognition: convergent patterns in the early evolution of learning. *Interface Focus* 7: 20160158.

Dunbar, R. I. M. (1995). Neocortex size and group size in primates: a test of the hypothesis. *Journal of Human Evolution* 28: 287–296.

Dunn, M., Greenhill, S. J., Levinson, S. C. and Gray, R. D. (2011). Evolved structure of language shows lineage-specific trends in word-order universals. *Nature* 473: 79–82.

Duntley, J. D. and Buss, D. M. (2011). Homicide adaptations. *Aggression and Violent Behavior* 16: 399–410.

Durham, W. H. (1991). *Coevolution: Genes, Culture and Human Diversity*. Stanford, CA: Stanford University Press.

Eagly, A. H. and Wood. W. (1999). The origins of sex differences in human behavior: evolved dispositions versus social roles. *American Psychologist* 54: 408–423.

Edelman, G. M. (1987). *Neural Darwinism: The Theory of Neurological Group Selection*. New York, NY: Basic Books.

Efferson, C., Lalive, R. and Fehr, E. (2008a). The coevolution of cultural groups and ingroup favoritism. *Science* 321: 1844–1849.

Efferson, C., Lalive, R., Richerson, P. J., McElreath, R. and Lubell, M. (2008b). Conformists and mavericks: the empirics of frequency-dependent cultural transmission. *Evolution and Human Behavior* 29: 56–64.

Eldredge, N., Pievani, T., Serrelli, E. and Temkin, I. (Eds.) (2016). *Evolutionary Theory: A Hierarchical Perspective*. Chicago, IL: University of Chicago Press.

Ellis, B. J., Del Guidice, M., Dishion, T. J., Figueredo, A. J., Gray, P., Griskevicius, V., Hawley, P. H., Jacobs, W. J., et al. (2012). The evolutionary basis of risky adolescent behavior: implications for science, policy, and practice. *Developmental Psychology* 48: 598–623.

Ellis, E. (2024). The Anthropocene condition: evolving through socio-ecological transformations. *Proceedings of the Royal Society B* 379: 20220255.

Ellis, E. C., Gauthier, N., Goldewijk, K. K., Bliege Bird, R., Boivin, N., Díaz, S., Fuller, D. Q., Gill, J. L., et al. (2021). People have shaped most of terrestrial nature for at least 12,000 years. *Proceedings of the National Academy of Sciences* 118: e2023483118.

Ellis, P. (Ed.) (2001). *Reproductive Ecology and Human Evolution*. New York, NY: Aldine de Gruyter.

Emery, N. J. (2004). Are corvids 'feathered apes'? Cognitive evolution in crows, jays, rooks and jackdaws. In S. Watanabe (Ed.) *Comparative Analysis of Minds* (pp. 181–213). Tokyo, Japan: Keio University Press.

Endler, J. A. (1986a). *Natural Selection in the Wild*. Princeton, NJ: Princeton University Press.
Endler, J. A. (1986b). The newer synthesis? Some conceptual problems in evolutionary biology. *Oxford Surveys in Evolutionary Biology* 3: 224–243.
Eriksson, K., Coultas, J. C. and de Barra, M. (2016). Cross-cultural differences in emotion selection on transmission of information. *Journal of Cognition and Culture* 16: 122–143.
Eriksson, K., Enquist, M. and Ghirlanda, S. (2007). Critical points in current theory of conformist social learning. *Journal of Evolutionary Psychology* 5: 67–87.
Evans, C. L., Greenhill, S. J., Watts, J., List, J., Botero, C. A., Gray, R. D. and Kirby, K. R. (2021). The uses and abuses of tree thinking in cultural evolution. *Philosophical Transactions of the Royal Society B* 376: 20200056.
Evans, R. I. (1975). *Konrad Lorenz: The Man and His Ideas*. New York, NY: Harcourt Brace.
Evershed, R. P., Davey Smith, G., Roffet-Salque, M., Timpson, A., Diekmann, Y., Lyon, M. S., Cramp, L. J. E., Casanova, E., et al. (2022). Dairying, diseases and the evolution of lactase persistence in Europe. *Nature* 608: 336–345.
Ewing, L., Rhodes, G. and Pellicano, E. (2010). Have you got the look? Graze direction affects judgements of facial attractiveness. *Visual Cognition* 18: 321–330.
Faith, J. T., Du, A., Behrensmeyer, A. K., Davies, B., Patterson, D. B., Rowan, J. and Wood, B. (2021). Rethinking ecological drivers of hominin evolution. *Trends in Ecology and Evolution* 36: 797–807.
Fan, S., Hansen, M. E. B., Lo, Y. and Tishkoff, S. A. (2016). Going global by adapting local: a review of recent human adaptation. *Science* 354: 54–59.
Fawcett, T. W., Hamblin, S. and Giraldeau, L. (2013). Exposing the behavioral gambit: the evolution of learning and decision rules. *Behavioral Ecology* 24: 2–11.
Fehr, E. and Fischbacher, U. (2003). The nature of human altruism. *Nature* 425: 785–791.
Feldman, M. W. and Cavalli-Sforza, L. L. (1976). Cultural and biological evolutionary processes, selection for a trait under complex transmission. *Theoretical Population Biology* 9: 238–259.
Feldman, M. W. and Cavalli-Sforza, L. L. (1989). On the theory of evolution under genetic and cultural transmission with application to the lactose absorption problem. In M. W. Feldman (Ed.) *Mathematical Evolutionary Theory* (pp. 145–173). Princeton, NJ: Princeton University Press.
Feldman, M. W. and Laland, K. N. (1996). Gene–culture coevolutionary theory. *Trends in Ecology and Evolution* 11: 453–457.
Feldman, M. W. and Lewontin, R. C. (1975). The heritability hang-up. *Science* 190: 1163–1168.
Feldman, M. W. and Lewontin, R. C. (2008). Race, ancestry, and medicine. In B. Koenig, S. S. Lee and S. Richardson (Eds.) *Revisiting Race in a Genomic Era* (pp. 89–101). New Brunswick, NJ: Rutgers University Press.
Feldman, M. W., Lewontin, R. C and King, M. (2003). Race: a genetic melting-pot. *Nature* 424: 374.
Feldman, M. W. and Ramachandran, S. (2018). Missing compared to what? Revisiting heritability, genes and culture. *Philosophical Transactions of the Royal Society B* 373: 20170064.
Fessler, D. M. T. and Haley, K. (2006). Guarding the perimeter: the outside-inside dichotomy in disgust and bodily experience. *Cognition and Emotion* 20: 3–19.
Fessler, D. M. T. and Machery, E. (2012). Culture and cognition. In: E. Margolis, R. Samuels and S. P. Stich (Eds.) *The Oxford Handbook of Philosophy of Cognitive Science* (pp. 503–527). Oxford, UK: Oxford University Press.
Fessler, D. M. T. and Navarrette, C. D. (2003). Domain-specific variation in disgust sensitivity across the menstrual cycle. *Evolution and Human Behavior* 24: 406–417.
Fiddick, L., Cosmides, L. and Tooby, J. (2000). No interpretation without representation: the role of domain-specific representations and inferences in the Wason selection task. *Cognition* 77: 1–79.

Firmin, A. (1885). *De l'Égalité des Races Humaines*. Paris, France: Libraire Cotillon (reprinted in English, 2002, Champaign, IL: University of Illinois Press).

Fisher, R. A. (1930). *Genetical Theory of Natural Selection*. Oxford, UK: Clarendon Press (reprinted 1999, Oxford, UK: Oxford University Press).

Fluehr-Lobban, C. (2000). Anténor Firmin: Haitian pioneer of anthropology. *American Anthropologist* **102**: 449–466.

Flynn, E. G., Laland, K. N., Kendal, R. L. and Kendal, J. R. (2013). Developmental niche construction. *Developmental Science* **16**: 296–313.

Fodor, J. A. (2000). *The Mind Doesn't Work That Way*. Cambridge, MA: MIT Press.

Fogarty, L., Creanza, N. and Feldman, M. W. (2015). Cultural evolutionary perspectives on creativity and human innovation. *Trends in Ecology and Evolution* **30**: 736–754.

Fogarty, L., Creanza, N. and Feldman, M. W. (2019). The life history of learning: demographic structure changes cultural outcomes. *PLoS Computational Biology* **15**: e1006821.

Foister, T. I. F., Žliobaitė, I., Wilson, O. E., Fortelius, M. and Tallavaara, M. (2023). Homo heterogenus: variability in early Pleistocene *Homo* environments. *Evolutionary Anthropology* **32**: 373–385.

Foley, R. (1995). The adaptive legacy of human evolution: a search for the environment of evolutionary adaptedness. *Evolutionary Anthropology* **4**: 194–203.

Forrest, D. W. (1974). *Francis Galton: The Life and Work of a Victorian Genius*. London, UK: Paul Elek.

Fortunato, L. and Archetti, M. (2010). Evolution of monogamous marriage by maximization of inclusive fitness. *Journal of Evolutionary Biology* **23**: 149–156.

Fouts, H. N. (2008). Father involvement with young children among the Aka and Bofi foragers. *Cross-Cultural Research* **42**: 290–312.

Fox Keller, E. (2010). *Mirage of a Space Between Nature and Nurture*. Durham, NC: Duke University Press.

Francis, M. (2007). *Herbert Spencer and the Invention of Modern Life*. Stocksfield, UK: Acumen.

Francois, P., Fujiwara, T. and van Ypersele, T. (2018). The origins of prosociality: cultural group selection in the workplace and the laboratory. *Science Advances* **4**: eaat2201.

Frankenhuis, W. E. and Amir, D. (2022). What is the expected human childhood? Insights from evolutionary anthropology. *Development and Psychopathology* **34**: 473–497.

Fuentes, A. (2021). 'The Descent of Man', 150 years on. *Science* **372**: 769.

Gao, Z. (2024). Unveiling recent and ongoing adaptive selection in human populations. *PLoS Biology* **22**: e3002469.

Galton, F. (1869). *Hereditary Genius*. London, UK: Julian Friedman Publishers.

Galton, F. (1883). *Inquiries into Human Faculty and its Development*. London, UK: Macmillan.

Galton, F. (1904). Eugenics: its definition, scope, and aims. *American Journal of Sociology* **10**: 1–25.

Gangestad, S. W., Haselton, M. G. and Buss, D. M. (2006). Evolutionary foundations of cultural variation: evoked culture and mate preferences. *Psychological Inquiry* **17**: 75–95.

Gangestad, S. W. and Simpson, J. A. (2000). The evolution of human mating: trade-offs and strategic pluralism. *Behavioral and Brain Sciences* **23**: 573–644.

Gangestad, S. W. and Simpson, J. A. (2007). Whither sciences of the evolution of mind? In S. W. Gangestad and J. A. Simpson (Eds.) *The Evolution of Mind: Fundamental Questions and Controversies* (pp. 397–437). New York, NY: Guilford Press.

Garcia, J. and Koelling, R. A. (1966). Prolonged relation of cue to consequence in avoidance learning. *Psychonomic Science* **4**: 123–124.

Gerbault, P., Liebert, A., Itan, Y., Powell, A., Currat, M., Burger, J., Swallow, D. M. and Thomas, M. G. (2011). Evolution of lactase persistence: an example of human niche construction. *Philosophical Transactions of the Royal Society B* **366**: 863–877.

Gibson, M. A. and Lawson, D. W. (2015). Applying evolutionary anthropology. *Evolutionary Anthropology* **24**: 3–14.

Gibson, M. A. and Mace, R. (2006). An energy-saving development initiative increases birth rate and childhood malnutrition in rural Ethiopia. *PLoS Medicine* **3**: e87.

Gibson, M. A. and Mace, R. (2007). Polygyny, reproductive success and child health in rural Ethiopia: why marry a married man? *Journal of Biosocial Sciences* **39**: 287–300.

Gintis, H. (2009). *Game Theory Evolving: A Problem-Centered Introduction to Modelling Strategic Interaction* (2nd edition). Princeton, NJ: Princeton University Press.

Goetz, C. D., Pillsworth, E. G., Buss, D. M. and Conroy-Beam, D. (2019). Evolutionary mismatch in mating. *Frontiers in Psychology* **10**: 2709.

Gokcumen, O. (2020). Archaic hominin introgression into modern human genomes. *Yearbook of Physical Anthropology* **171** (Supplement 70): 60–73.

Goodall, J. (1986). *The Chimpanzees of Gombe: Patterns of Behavior*. Cambridge, MA: Harvard University Press.

Goodman, A., Koupil, I. and Lawson, D. W. (2012). Low fertility increases descendant socioeconomic position but reduces long-term fitness in a modern post-industrial society. *Proceedings of the Royal Society B* **279**: 4342–4351.

Gottlieb, G. (1971). *Development of Species Identification in Birds*. Chicago, IL: University of Chicago Press.

Gould, S. J. (1980). *The Panda's Thumb*. Middlesex, UK: Penguin.

Gould, S. J. (1985). *Ontogeny and Phylogeny*. Harvard, MA: Harvard University Press.

Gould, S. J. (1991). *Bully for Brontosaurus: Reflections in Natural History*. New York, NY: Norton.

Gould, S. J. and Vrba, E. (1982). Exaptation: a missing term in the science of form. *Paleobiology* **8**: 4–15.

Grafen, A. (1984). Natural selection, kin selection and group selection. In J. Krebs and N. B. Davies (Eds.) *Behavioural Ecology: An Evolutionary Approach* (2nd edition) (pp. 62–84). Oxford, UK: Blackwell Scientific Publications.

Graves, J. L. and Goodman, A. H. (2021). *Racism Not Race: Answers to Frequently Asked Questions*. New York, NY: Columbia University Press.

Gray, R. D. and Jordan, F. M. (2000). Language trees support the express-train sequence of Austronesian expansion. *Nature* **405**: 1052–1055.

Greenhill, S. J., Currie, T. E. and Gray, R. D. (2009). Does horizontal transmission invalidate cultural phylogenies? *Proceedings of the Royal Society B* **276**: 2299–2306.

Gruber, H. E. (1974). *Darwin on Man*. New York, NY: Dutton.

Guglielmino, C. R., Viganotti, C., Hewlett, B. and Cavalli-Sforza, L. L. (1995). Cultural variation in Africa: role of mechanism of transmission and adaptation. *Proceedings of the National Academy of Sciences* **92**: 7585–7589.

Gurven, M. (2004). To give and to give not: the behavioral ecology of human food transfers. *Behavioral and Brain Sciences* **27**: 543–583.

Gurven, M., Stieglitz, J., Trumble, B., Blackwell, A. D., Beheim, B., Davis, H., Hooper, P. and Kaplan, H. (2017). The Tsimane Health and Life History Project: integrating anthropology and biomedicine. *Evolutionary Anthropology* **26**: 54–73.

Hackman, J. and Kramer, K. L. (2022). Kin networks and opportunities for reproductive cooperation and conflict among hunter-gatherers. *Philosophical Transactions of the Royal Society B* **378**: 20210434.

Haeckel, E. (1866). *Generelle Morphologie der Organismen* (Volumes I and II). Berlin, Germany: Georg Reimer.

Hahn, M. W. and Bentley, R. A. (2003). Drift as a mechanism for cultural change: an example from baby names. *Proceedings of the Royal Society B* **270**: S120–S123.

Hahn, M. and Xu, Y. (2022). Crosslinguistic word order variation reflects evolutionary pressures of dependency and information locality. *Proceedings of the National Academy of Sciences* 119: e2122604119.

Haig, D. (1993). Genetic conflicts in human pregnancy. *Quarterly Review of Biology* 68: 495–532.

Haldane, J. B. S. (1955). Population genetics. *New Biology* 18: 34–51.

Hall, G. S. (1904). *Adolescence: Its Psychology and its Relation to Physiology, Anthropology, Sociology, Sex, Crime, Religion and Education* (Volumes 1 and 2). New York, NY: Appleton and Company.

Hamilton, W. (1964). The genetical evolution of social behaviour: I. *Journal of Theoretical Biology* 7: 1–16.

Hamilton, W. (1964). The genetical evolution of social behaviour: II. *Journal of Theoretical Biology* 7: 17–32.

Handley, C. and Mathew, S. (2020). Human large-scale cooperation as a product of competition between cultural groups. *Nature Communications* 11: 702.

Hare, B. (2017). Survival of the friendliest: *Homo sapiens* evolved via selection for prosociality. *Annual Review of Psychology* 68: 155–186.

Harris, M. (2001). *The Rise of Anthropological Theory: A History of Theories of Culture*. Walnut Creek, CA: AltaMira Press.

Harvati, K. and Reyes-Centeno, H. (2022). Evolution of *Homo* in the Middle and Late Pleistocene. *Journal of Human Evolution* 173: 103279.

Hawkes, K. (1991). Showing off: tests of another hypothesis about men's foraging goals. *Ethology and Sociobiology* 11: 29–54.

Hawkes, K. (2003). Grandmothers and the evolution of human longevity. *American Journal of Human Biology* 15: 280–400.

Hawkes, K., O'Connell, J. F., and Blurton Jones, N. G. (1989). Hardworking Hadza grandmothers. In V. Standen and R. A. Foley (Eds.) *Comparative Socioecology: The Behavioural Ecology of Humans and Other Mammals* (pp. 341–366). Oxford, UK: Blackwell Scientific Publications.

Hayward, A. D., Richard, I. J. and Lummaa, V. (2013). Influence of early-life nutrition on mortality and reproductive success during a subsequent famine in a preindustrial population. *Proceedings of the National Academy of Sciences* 110: 13886–13891.

Heath, C., Bell, C. and Sternberg, E. (2001). Emotional selection in memes: the case of urban legends. *Journal of Personality and Social Psychology* 81: 1028–1041.

Heggarty, P., Anderson, C., Scarborough, M., King, B., Bouckaert, R., Jocz, L., Kümmel, M. J., Jügel, T., et al. (2023). Language trees with sampled ancestors support a hybrid model for the origin of Indo-European languages. *Science* 381: eabg0818.

de Heij, M. E., van den Hout and Tinbergen, J. M. (2006). Fitness cost of incubation in great tits (*Parus major*) is related to clutch size. *Proceedings of the Royal Society B* 273: 2353–2361.

Heine, S. J. (2020). *Cultural Psychology* (4th edition). New York, NY: Norton.

Henrich, J. (2004). Demography and cultural evolution: how adaptive cultural processes can produce Maladaptive losses: the Tasmanian case. *American Antiquity* 69: 197–214.

Henrich, J. (2016). *The Secret of Our Success: How Culture is Driving Human Evolution, Domesticating our Species, and Making Us Smarter*. Princeton, NJ: Princeton University Press.

Henrich, J. and Boyd, K. (1998). The evolution of conformist transmission and the emergence of between-group differences. *Evolution and Human Behavior* 19: 215–242.

Henrich, J. and Boyd, K. (2002). On modeling cognition and culture: why cultural evolution does not require replication of representations. *Journal of Cognition and Culture* 2: 87–112.

Henrich, J., Boyd, R., Bowles, S., Camerer, C., Fehr, E. and Gintis, H. (Eds.) (2004). *Foundations of Human Sociality*. Oxford, UK: Oxford University Press.

Henrich, J., Boyd, R., Bowles, S., Camerer, C., Fehr, E., Gintis, H., and McElreath, R. (2001). In search of *Homo economicus*: behavioral experiments in 15 small-scale societies. *American Economic Review* **91**: 73–77.

Henrich, J., Boyd, R. and Richerson, P. J. (2008). Five misunderstandings about cultural evolution. *Human Nature* **19**: 119–137.

Henrich, J., Heine, S. J. and Norenzayan, A. (2010). The weirdest people in the world? *Behavioral and Brain Sciences* **33**: 61–135.

Henrich, J. and McElreath, R. (2003). The evolution of cultural evolution. *Evolutionary Anthropology* **12**: 123–135.

Henrich, J. and Muthukrishna, M. (2021). The origins and psychology of human cooperation. *Annual Review of Psychology* **72**: 207–240.

Herculano-Houzel, S. (2020). Remarkable, but not special: what human brains are made of. In J. H. Kaas (Ed.) *Evolutionary Neuroscience* (2nd edition) (pp. 803–813). Cambridge, MA: Academic Press.

Herzog, H. A., Bentley, R. A. and Hahn, M. W. (2004). Random drift and large shifts in popularity of dog breeds. *Proceedings of the Royal Society B* **271**: S353–S356.

Heyes, C. (2000). Evolutionary psychology in the round. In C. Heyes and L. Huber (Eds.) *The Evolution of Cognition* (pp. 3–22). Cambridge, MA: MIT Press.

Heyes, C. (2012). Grist and mills: on the cultural origins of cultural learning. *Philosophical Transactions of the Royal Society B* **367**: 2181–2191.

Heyes, C. (2018). *Cognitive Gadgets: The Cultural Evolution of Thinking*. Oxford, UK: Oxford University Press.

Heyes, C., Bang, D., Shea, N., Frith, C. D. and Fleming, S. M. (2020). Knowing ourselves together: the cultural origins of metacognition. *Trends in Cognitive Sciences* **24**: 349–362.

Hill, K. (1988). Macronutrient modifications of optimal foraging theory: an approach using indifference curves applied to some modern foragers. *Human Ecology* **16**: 157–197.

Hill, K. and Hurtado, A. M. (1996). *Ache Life History: The Ecology and Demography of a Foraging People*. New York, NY: Aldine de Gruyter.

Hinde, R. A. (1956). Ethological models and the concept of 'drive'. *British Journal of the Philosophy of Science* **6**: 321–331.

Hinde, R. A. (1966). *Animal Behaviour: A Synthesis of Ethology and Comparative Psychology*. New York, NY: McGraw-Hill.

Hinde, R. A. (1974). *Biological Bases of Human Social Behaviour*. New York, NY: McGraw-Hill.

Hinde, R. A. (1982). *Ethology*. Glasgow, UK: Fontana Press.

Hinde, R. A. (1987). *Individuals, Relationships and Culture*. Cambridge, UK: Cambridge University Press.

Hinde, R. A. and Barden, L. A. (1985). The evolution of the teddy bear. *Animal Behavior* **33**: 1371–1373.

Hochman, A. (2021). Janus-faced race: is race biological, social, or mythical? *American Journal of Physical Anthropology* **175**: 453–464.

Holden, C. J. and Mace, R. (2003). Spread of cattle led to the loss of matrilineal descent in Africa: a coevolutionary analysis. *Proceedings of the Royal Society B* **270**: 2425–2433.

Holekamp, K. E., Swanson, E. M. and Van Meter, P. E. (2013). Developmental constraints on behavioural flexibility. *Philosophical Transactions of the Royal Society B* **368**: 20120350.

Hopcroft, R. L. (2006). Sex, status and reproductive success in the contemporary United States. *Evolution and Human Behavior* **27**: 104–120.

Hoppitt, W. J. E., Brown, G. R., Kendal, R., Rendell, L., Thornton, A., Webster, M. M. and Laland, K. N. (2008). Lessons from animal teaching. *Trends in Ecology and Evolution* **23**: 486–493.

Horner, V. and Whiten, A. (2005). Causal knowledge and imitation/emulation switching in chimpanzees (*Pan troglodytes*) and children (*Homo sapiens*). *Animal Cognition* **8**: 164–181.

Hout, M., Greeley, A. and Wilde, M. J. (2001). The demographic imperative in religious change in the United States. *American Journal of Sociology* **107**: 468–500.

Hrdy, S. B. (1977). *The Langurs of Abu: Female and Male Strategies of Reproduction*. Cambridge, MA: Harvard University Press.

Hrdy, S. B. (1981). *The Woman that Never Evolved*. Cambridge, MA: Harvard University Press.

Hrdy, S. B. (1999). *Mother Nature: Natural Selection and the Female of the Species*. London, UK: Chatto and Windus.

Hrdy, S. B. (2009). *Mothers and Others: The Evolutionary Origins of Mutual Understanding*. Cambridge, MA: Harvard University Press.

Hubel, D. H. and Weisel, T. N. (2005). *Brain and Visual Perception: The Story of a 25-Year Collaboration*. Oxford, UK: Oxford University Press.

Hull, D. L. (1982). The naked meme. In H. C. Plotkin (Ed.) *Learning, Development, and Culture* (pp. 273–327). Chichester, UK: Wiley.

Huxley, J. S. (1942). *Evolution: The Modern Synthesis*. London, UK: Allen and Unwin.

Huxley, T. H. (1863). *Evidence as to Man's Place in Nature*. London, UK: Williams and Norgate.

Ingold, T. (2007). The trouble with 'evolutionary biology'. *Anthropology Today* **23**: 13–17.

Jablonka, E. and Lamb, M. J. (2014). *Evolution in Four Dimensions: Genetic, Epigenetic, Behavioral, and Symbolic Variation in the History of Life* (revised edition). Cambridge, MA: MIT Press.

Jaeggi, A. V. and Gurven, M. (2013). Reciprocity explains food sharing in humans and other primates independent of kin selection and tolerated scrounging: a phylogenetic meta-analysis. *Proceedings of the Royal Society B* **280**: 20131615.

James, W. (1890). *Principles of Psychology*. New York, NY: Holt.

Jang, H., Janmaat, K. R. L., Kandza, V. and Boyette, A. H. (2022). Girls in early childhood increase food returns of nursing women during subsistence activities of the BaYaka in the Republic of Congo. *Proceedings of the Royal Society B* **289**: 20221407.

Jang, H., Ross, C. T., Boyette, A. H., Janmaat, K. R. L., Kandza, V. and Redhead, D. (2024). Women's subsistence networks scaffold cultural transmission among BaYaka foragers in the Congo Basin. *Sciences Advances* **10**: eadj2543.

Jasienska, G. (2013). *The Fragile Wisdom: An Evolutionary View on Women's Biology and Health*. Cambridge, MA: Harvard University Press.

Jensen, A. (1969). How much can we boost IQ and scholastic achievement? *Harvard Educational Review* **39**: 1–123.

Jiang, Q. and Zhang, C. (2021). Recent sex ratio at birth in China. *BMJ Global Health* **6**: e005438.

Johnson, M., Henriksen, R. and Wright, D. (2021). The neural crest cell hypothesis: no unified explanation for domestication. *Genetics* **219**: iyab097.

Kalikow, T. J. (2020). Konrad Lorenz on human degeneration and social decline: a chronic preoccupation. *Animal Behavior* **163**: 267–272.

Kameda, T. and Nakanishi, D. (2002). Cost-benefit analysis of social/cultural learning in a nonstationary uncertain environment: an evolutionary simulation and an experiment with human subjects. *Evolution and Human Behavior* **23**: 373–393.

Kameda, T. and Nakanishi, D. (2003). Does social/cultural learning increase human adaptability? Roger's question revisited. *Evolution and Human Behavior* **24**: 242–260.

Kameda, T., Toyokawa, W. and Tindale, R. S. (2022). Information aggregation and collective intelligence beyond the wisdom of crowds. *Nature Reviews Psychology* **1**: 345–357.

Kant, I. (1781). *Critique of Pure Reason* (reprinted 1993, London, UK: Everyman).

Kaplan, H. S. and Gangestad, S. W. (2005). Life history theory and evolutionary psychology. In D. M. Buss (Ed.) *The Handbook of Evolutionary Psychology* (pp. 68–96). Hoboken, NJ: Wiley.

Kaplan, H. and Hill, K. (1985). Hunting ability and reproductive success among male Ache foragers. *Current Anthropology* **26**: 131–133.

Kaplan, H. S. and Lancaster, J. B. (2000). The evolutionary economics and psychology of the demographic transition to low fertility. In L. Cronk, N. Chagnon and W. Irons (Eds.) *Adaptation and Human Behavior: An Anthropological Perspective* (pp. 283–322). New York, NY: Aldine de Gruyter.

Kaplan, H., Hill, K., Lancaster, J. and Hurtado, A. M. (2000). A theory of human life history evolution: diet, intelligence, and longevity. *Evolutionary Anthropology* 9: 156–185.

Kappeler, P. M., Barrett, L., Blumstein, D. T. and Clutton-Brock, T. H. (2013). Constraints and flexibility in mammalian social behaviour: introduction and synthesis. *Philosophical Transactions of the Royal Society B* 368: 20120337.

Karmiloff-Smith, A. (2000). Why babies' minds aren't Swiss Army Knives. In Rose, H. and Rose, S. (Eds.) *Alas Poor Darwin: Arguments against Evolutionary Psychology* (pp. 144–156). London, UK: Cape.

Kaufman, A. B., Call, J. and Kaufman, J. C. (Eds.) (2021). *The Cambridge Handbook of Animal Cognition*. Cambridge, UK: Cambridge University Press.

Kelley, J. L. and Swanson, W. J. (2008). Dietary change and adaptive evolution of enamelin in humans and among primates. *Genetics* 178: 1595–1603.

Kelly, R. L. (2013). *The Lifeways of Hunter-Gatherers: The Foraging Spectrum*. Cambridge, UK: Cambridge University Press.

Kendal, R. L., Boogert, N. J., Rendell, L., Laland, K. N., Webster, M. and Jones, P. L. (2018). Social learning strategies: bridge-building between fields. *Trends in Cognitive Sciences* 22: 651–665.

Kennett, D. J., Breitenbach, S. F. M., Aquino, V. V., Asmerom, Y., Awe, J., Baldini, J. U. L., Bartlein, P., Culleton, B. J., et al. (2012). Development and disintegration of Maya political systems in response to climate change. *Science* 338: 788–791.

Khaitovich, P., Enard, W., Lachmann, M. and Pääbo, S. (2006). Evolution of primate gene expression. *Nature Reviews Genetics* 7: 693–702.

Kim, H. S. and Sasaki, J. Y. (2014). Cultural neuroscience: biology of the mind in cultural contexts. *Annual Review of Psychology* 65: 487–514.

King, A. J. and Marshall, H. H. (2022). Optimal foraging. *Current Biology* 32: R680–R683.

King, C. (2019). *The Reinvention of Humanity: A Story of Race, Sex, Gender and the Discovery of Culture*. London, UK: Vintage.

Kingsolver, J. G., Hoekstra, H. E., Hoekstra, J. M., Berrigan, D., Vignieri, S. N., Hill, C. E., Hoang, A., Gilbert, P., et al. (2001). The strength of phenotypic selection in natural populations. *American Naturalist* 157: 245–261.

Kirby, S. Cornish, H. and Smith, K. (2008). Cumulative cultural evolution in the laboratory: an experimental approach to the origins of structure in human language. *Proceedings of the National Academy of Sciences* 105: 10681–10686.

Kirby, S., Tamariz, M., Cornish, H. and Smith, K. (2015). Compression and communication in the cultural evolution of linguistic structure. *Cognition* 141: 87–102.

Kirkpatrick, R. C. (2000). The evolution of human homosexual behavior. *Current Anthropology* 41: 385–413.

Kitcher P. (1985). *Vaulting Ambition. Sociobiology and the Quest for Human Nature*. Cambridge, MA: MIT Press.

Klenerman, P. (2017). *The Immune System: A Very Short Introduction*. Oxford, UK: Oxford University Press.

Kolodny, O. and Edelman, S. (2018). The evolution of the capacity for language: the ecological context and adaptive value of a process of cognitive hijacking. *Philosophical Transactions of the Royal Society B* 373: 20170052.

Koster, J. M. (2008). Hunting with dogs in Nicaragua: an optimal foraging approach. *Current Anthropology* 49: 935–944.

Koster, J. and Bird, D. (2024). Foraging strategies. In J. Koster, B. Scelza and M. K. Shenk (Eds.) *Human Behavioral Ecology* (pp. 48–75). Cambridge, UK: Cambridge University Press.

Koster, J., Scelza, B. and Shenk, M. K. (Eds.) (2024). *Human Behavioral Ecology*. Cambridge, UK: Cambridge University Press.

Kramer, K. L. and Veile, A. (2018). Infant allocare in traditional societies. *Physiology and Behavior* 193: 117–126.

Kruger, D. J. and Nesse, R. M. (2006). An evolutionary life-history framework for understanding sex differences in human mortality rates. *Human Nature* 17: 74–97.

Kruuk, H. (2004). *Niko's Nature*. Oxford, UK: Oxford University Press.

Kuhlwilm, M. and Boeckx, C. (2019). A catalog of single nucleotide changes distinguishing modern humans from archaic hominins. *Scientific Reports* 9: 8463.

Kumm, J., Laland, K. N., and Feldman, M. W. (1994). Gene–culture coevolution and sex ratios: the effects of infanticide, sex-selective abortion, and sex-biased parental investment on the evolution of sex ratios. *Theoretical Population Biology* 46: 249–278.

Kuper, A. (1994). *The Chosen Primate*. Cambridge, MA: Harvard University Press.

Lack, D. (1954). *The Natural Regulation of Animal Numbers*. Oxford, UK: Oxford University Press.

Lack, D. (1966). *Population Studies of Birds*. Oxford, UK: Oxford University Press.

Lacy, S. and Ocobock, C. (2024). Woman the hunter: the archaeological evidence. *American Anthropologist* 126: 19–31.

Lala, K., Uller, T., Feiner, N., Feldman, M. W. and Gilbert, S. F. (2024). *Evolution Evolving: The Developmental Origins of Adaptation and Biodiversity*. Princeton, NJ: Princeton University Press.

Laland, K. N. (2017). *Darwin's Unfinished Symphony: How Culture Made the Human Mind*. Princeton, NJ: Princeton University Press.

Laland, K. N. and Brown, G. R. (2002). *Sense and Nonsense* (1st edition). Oxford, UK: Oxford University Press.

Laland, K. N. and Brown, G. R. (2006). Niche construction, human behaviour and the adaptive-lag hypothesis. *Evolutionary Anthropology* 15: 95–104.

Laland, K. N. and Brown, G. R. (2008). Commentary of 'The chimpanzee has no clothes' by Sayers and Lovejoy. *Current Anthropology* 49: 101–102.

Laland, K. N. and Brown, G. R. (2018). The social construction of human nature. In E. Hannon and T. Lewens (Eds.) *Why We Disagree About Human Nature* (pp. 127–144). Oxford, UK: Oxford University Press.

Laland, K. N. and Galef, B. G. (Eds.) (2009). *The Question of Animal Culture*. Cambridge, MA: Harvard University Press.

Laland, K. N. and Odling-Smee, J. (2000). The evolution of the meme. In R. Aunger (Ed.) *Darwinizing Culture: The Status of Memetics as a Science* (pp. 121–141). Oxford, UK: Oxford University Press.

Laland, K., Odling-Smee, J. and Endler, J. (2017). Niche construction, sources of selection and trait covariation. *Interface Focus* 7: 20160147.

Laland, K. N., Odling-Smee, J., and Feldman, M. W. (2000). Niche construction, biological evolution, and cultural change. *Behavioral and Brain Sciences* 23: 131–175.

Laland, K. N., Odling-Smee, F. J. and Feldman, M. W. (2001). Cultural niche construction and human evolution. *Journal of Evolutionary Biology* 14: 22–33.

Laland, K. N., Odling-Smee, F. J. and Myles, S. (2010). How culture has shaped the human genome: bringing genetics and the human sciences together. *Nature Reviews Genetics* 11: 137–148.

Laland, K. N. and Seed, A. M. (2021). Understanding human cognitive uniqueness. *Annual Review of Psychology* 72: 689–716.

Laland, K. N., Sterelny, K., Odling-Smee, J., Hoppitt, W. and Uller, T. (2011). Cause and effect in biology revisited: is Mayr's proximate-ultimate dichotomy still useful? *Science* 334: 1512–1516.

Laland, K. N., Uller, T., Feldman, M., Sterelny, K., Müller, G. B., Moczek, A., Jablonka, E. and Odling-Smee, J. (2014). Does evolutionary theory need a rethink? Yes, urgently. *Nature* 514: 161–164.

Laland, K. N., Uller, T., Feldman, M. W., Sterelny, K., Müller, G. B., Moczek, A., Jablonka, E. and Odling-Smee, J. (2015). The extended evolutionary synthesis: its structure, assumptions and predictions. *Proceedings of the Royal Society B* 282: 20151019.

Lamarck, J. (1809). *Philosophie Zoologique, ou Exposition des Considérations relatives à l'histoire naturelle des Animaux.* Paris: Muséum Nationale d'Histoire Naturelles (reprinted 2011, Cambridge, UK: Cambridge University Press, Volumes 1 and 2).

Lambert, B., Kontonatsios, G., Mauch, M., Kokkoris, T., Jockers, M., Ananiadou, S. and Leroi, A. M. (2020). The pace of modern culture. *Nature Human Behaviour* 4: 352–360.

Lang, M., Purzychi, B. G., Apicella, C. L., Atkinson, Q. D., Bolyanatz, A., Cohen, E., Handley, C., Klocová, E. K., et al. (2019). Moralizing gods, impartiality and religious parochialism across 15 societies. *Proceedings of the Royal Society B* 286: 20190202.

Lawson, D. W. and Borgerhoff Mulder, M. (2016). The offspring quantity-quality trade-off and human fertility variation. *Philosophical Transactions of the Royal Society B* 371: 20150145.

Lawson, D. W. and Gibson, M. A. (2018). Polygynous marriage and child health in sub-Saharan Africa: what is the evidence of harm? *Demographic Research* 39: 177–208.

Lawson, D. and Mace, R. (2011). Parental investment and the optimisation of human family size. *Philosophical Transactions of the Royal Society B* 366: 333–343.

Leach, E. (1981). Biology and social science: wedding or rape? *Nature* 291: 267–268.

Lee, M., Lindo, J. and Rilling, J. K. (2022). Exploring gene-culture coevolution in humans by inferring neuroendophenotypes: a case study of the oxytocin receptor gene and cultural tightness. *Genes, Brain and Behavior* 21: e12783.

Lehrman, D. S. (1953). A critique of Konrad Lorenz's theory of instinctive behaviour. *Quarterly Review of Biology* 28: 337–363.

Lehrman, D. S. (1965). Interaction between internal and external environments in the regulation of the reproductive cycle of the ring dove. In F. A. Beach (Ed.) *Sex and Behavior* (pp. 355–380). New York, NY: Wiley.

Lewens, T. (2015). *Cultural Evolution.* Oxford, UK: Oxford University Press.

Lew-Levy, S., Lavi, N., Reckin, R., Cristobal-Azkarate, J. and Ellis-Davies, K. (2018). How do hunter-gatherer children learn social norms? A meta-ethnographic review. *Cross-Cultural Research* 52: 213–255.

Lew-Levy, S., Reckin, R., Lavi, N., Cristobal-Azkarate, J. and Ellis-Davies, K. (2017). How do hunter-gatherer children learn subsistence skills? A meta-ethnographic review. *Human Nature* 28: 367–394.

Lewontin, R. C. (1972). The apportionment of human diversity. In T. Donzhansky, M. K. Hecht and W. C. Steere (Eds.) *Evolutionary Biology* (Volume 6) (pp. 381–398). New York, NY: Springer.

Lewontin, R. C. (1977). Caricature of Darwinism. *Nature* 266: 284–284.

Lewontin, R. C. (1983). Gene, organism and environment. In D. S. Bendall (Ed.) *Evolution From Molecules to Men* (pp. 273–285). Cambridge, UK: Cambridge University Press.

Lewontin, R. C. (1991). *Biology as Ideology: The Doctrine of DNA.* Toronto, Canada: Anasi.

Lewontin, R. (2000). *The Triple Helix: Gene, Organism, and Environment.* Cambridge, MA: Harvard University Press.

Li, N., Feldman, M. W. and Li, S. (2000). Cultural transmission in a demographic study of sex ratio at birth in China's future. *Theoretical Population Biology* 58: 161–172.

Li, N. P., van Vugt, M. and Colarelli, S. M. (2018). The evolutionary mismatch hypothesis: implications for psychological science. *Current Directions in Psychological Science* **27**: 38–44.

Lickliter, R. and Honeycutt, H. (2013). A developmental evolutionary framework for psychology. *Review of General Psychology* **17**: 184–189.

Liddle, J. R., Shackelford, T. K. and Weekes-Shackelford, V. A. (2012). Why can't we all just get along? Evolutionary perspectives on violence, homicide, and war. *Review of General Psychology* **16**: 24–36.

Lieberman, D., Tooby, J. and Cosmides, L. (2007). The architecture of human kin detection. *Nature* **445**: 727–731.

Liu, J., Mosti, F. and Silver, D. L. (2021). Human brain evolution: emerging roles for regulatory DNA and RNA. *Current Opinion in Neurobiology* **71**: 170–177.

Lloyd, E. A. and Feldman, M. W. (2002). Evolutionary psychology: a view from evolutionary biology. *Psychological Inquiry* **13**: 150–156.

López Herráez, D., Bauchet, M., Tang, K., Theunert, C., Pugach, I., Li, J., Nandineni, M. R., Gross, A., et al. (2009). Genetic variation and recent positive selection in worldwide human populations: evidence from nearly 1 million SNPs. *PLoS One* **4**: e7888.

Lorenz, K. (1950). The comparative method in studying innate behaviour patterns. *Symposium of the Society of Experimental Biology* **4**: 221–268.

Lorenz, K. (1965). *Evolution and Modification of Behavior*. Chicago, IL: University of Chicago Press.

Lorenz, K. (1966). *On Aggression*. London, UK: Methuen (reprinted 1996, Routledge).

Lotem, A., Kolodny, O. and Arbilly, M. (2023). Gene–culture coevolution in the cognitive domain. In J. Tehrani, J. Kendal and R. Kendal (Eds.) *The Oxford Handbook of Cultural Evolution*. Oxford, UK: Oxford University Press. https://doi.org/10.1093/oxfordhb/9780198869252.013.66

Luca, F., Perry, G. H. and Di Rienzo, A. (2010). Evolutionary adaptations to dietary changes. *Annual Review of Nutrition* **30**: 291–314.

Lukas, D. and Huchard, E. (2014). The evolution of infanticide by males in mammalian societies. *Science* **346**: 841–844.

Lumsden, C. J. and Wilson, E. O. (1981). *Genes, Mind, and Culture: The Coevolutionary Process*. Cambridge, MA: Harvard University Press.

Lynch, A. (1996). *Thought Contagion: How Belief Spreads through Society*. New York, NY: Basic Books.

Lynch, M. (2007). The frailty of adaptive hypotheses for the origins of organismal complexity. *Proceedings of the National Academy of Sciences* **104**: 8597–8604.

Mace, R. (1996). When to have another baby: a dynamic model of reproductive decision-making and evidence from Gabbra pastoralists. *Ethology and Sociobiology* **17**: 263–273.

Mace, R. (2014). Human behavioral ecology and its evil twin. *Behavioral Ecology* **25**: 443–449.

Mace, R. and Jordan, F. M. (2011). Macro-evolutionary studies of cultural diversity: a review of empirical studies of cultural transmission and cultural adaptation. *Philosophical Transactions of the Royal Society B* **366**: 402–411.

Machery, E. (2008). A plea for human nature. *Philosophical Psychology* **21**: 321–329.

Maglo, K. N., Mersha, T. B. and Martin, L. J. (2016). Population genomics and the statistical values of race: an interdisciplinary perspective on the biological classification of human populations and implications for clinical genetic epidemiological research. *Frontiers in Genetics* **7**: 22.

Mameli, M. and Bateson, P. (2011). An evaluation of the concept of innateness. *Philosophical Transactions of the Royal Society B* **366**: 436–443.

Mangel, M. and Clark, C. W. (1988). *Dynamic Modeling in Behavioral Ecology*. Princeton, NJ: Princeton University Press.

Maricic, T., Gunther, V., Georgiev, O., Gehre, S., Ćurlin, M., Schreiweis, C., Naumann, R., Burbano, H. A., et al. (2013). A recent evolutionary change affects a regulatory element in the human *FOXP2* gene. *Molecular Biology and Evolution* **30**: 844–852.

Markov, A. V. and Markov. M. A. (2020). Runaway brain-culture coevolution as a reason for larger brains: exploring the 'cultural drive' hypothesis by computer modeling. *Ecology and Evolution* **10**: 6059–6077.

Marlowe, F. (2010). *Hadza: Hunter-Gatherers of Tanzania*. Berkeley, CA: University of California Press.

Marr, D. (1982). *Vision: A Computational Investigation into the Human Representation and Processing of Visual Information*. San Francisco, CA: Freeman.

Marshall, J. A. R. (2015). *Social Evolution Theory and Inclusive Fitness Theory: An Introduction*. Princeton, NJ: Princeton University Press.

Mata, R., Josef, A. K. and Hertwig, R. (2016). Propensity for risk taking across the life span and around the globe. *Psychological Science* **27**: 231–243.

Mathew, S. and Perreault, C. (2015). Behavioural variation in 172 small-scale societies indicates that social learning is the main mode of human adaptation. *Proceedings of the Royal Society B* **282**: 20150061.

Mathieson, I. and Terhorst, J. (2022). Direct detection of natural selection in Bronze Age Britain. *Genome Research* **32**: 2057–2067.

Matthews, B., De Meester, L., Jones, C. G., Ibelings, B. W., Bouma, T. J., Nuutinen, V., van de Koppel, J. and Odling-Smee, J. (2014). Under niche construction: an operational bridge between ecology, evolution, and ecosystem science. *Ecological Monographs* **84**: 245–263.

Mattison, S. M., Quinlan, R. J. and Hare, D. (2019). The expendable male hypothesis. *Philosophical Transactions of the Royal Society B* **374**: 20180080.

Mattison, S. M. and Sear, R. (2016). Modernizing evolutionary anthropology. *Human Nature* **27**: 335–350.

Matzig, D. N., Schmid, C. and Riede, F. (2023). Mapping the field of cultural evolutionary theory and methods in archaeology using bibliometric methods. *Humanities and Social Sciences Communications* **10**: 271.

Maynard Smith, J. (1964). Group selection and kin selection. *Nature* **201**: 1145–1147.

Maynard Smith, J. (1975). Survival through suicide. *New Scientist* **28**: 496–497.

Maynard Smith, J. (1978). Optimization theory in evolution. *Annual Review of Ecology and Systematics* **9**: 31–56.

Maynard Smith, J. (1982). *Evolution and the Theory of Games*. Cambridge, UK: Cambridge University Press.

Maynard Smith, J. and Price, G. (1973). The logic of animal conflict. *Nature* **246**: 15–18.

Maynard Smith, J. and Warren, N. (1982). Models of cultural and genetic change. *Evolution* **36**: 620–627.

Mayr, E. (1942). *Systematics and the Origin of Species*. New York, NY: Columbia University Press.

Mayr, E. (1961). Cause and effect in biology. *Science* **134**: 1501–1506.

Mayr, E. (1982). *The Growth of Biological Thought: Diversity, Evolution, and Inheritance*. Cambridge, MA: Cambridge University Press.

McCorduck, P. (2004). *Machines Who Think: A Personal Inquiry Into the History and Prospects of Artificial Intelligence*. Natick, MA: AK Peters.

McCullough, M. E., Kimeldorf, M. B. and Cohen, A. D. (2008). An adaptation for altruism? The social causes, social effects, and social evolution of gratitude. *Current Directions in Psychological Science* **17**: 281–285.

McDonald, M. M., Navarrete, C. D. and Van Vugt, M. (2012). Evolution and the psychology of intergroup conflict: the male warrior hypothesis. *Philosophical Transactions of the Royal Society B* **367**: 670–679.

McDougall, W. (1908). *An Introduction to Social Psychology*. Boston, MA: J. W. Luce & Co.

McElreath, R. and Boyd, R. (2007). *Mathematical Models of Social Evolution: A Guide for the Perplexed*. Chicago, IL: University of Chicago Press.

McElreath, R. and Koster, J. (2024). The ends of human behavioral ecology. In J. Koster, B. Scelza and M. K. Shenk (Eds.) *Human Behavioral Ecology* (pp. 402-419). Cambridge, UK: Cambridge University Press.

McElreath, R., Lubell, M., Richerson, P. J., Waring, T. M., Baum, W. Edsten, E. Efferson, C. and Paciotti, B. (2005). Applying evolutionary models to the laboratory study of social learning. *Evolution and Human Behavior* 26: 483-508.

McElreath, R. and Strimling, P. (2008). When natural selection favors imitation of parents. *Current Anthropology* 49: 307-316.

McLain, D. K., Setters, D., Moulton, M. P. and Pratt, A. E. (2000). Ascription of resemblance of newborns by parents and nonrelatives. *Evolution and Human Behavior* 21: 11-23.

McNamara, J. M. and Houston, A. I. (2009). Integrating function and mechanism. *Trends in Ecology and Evolution* 24: 670-675.

Mead, M. (1928). *Coming of Age in Samoa: A Psychological Study of Primitive Youth for Western Civilization*. New York, NY: William Morrow and Company (reprinted 2001, New York, NY: HarperCollins).

Mead, M. (1930). *Growing up in New Guinea: A Comparative Study of Primitive Education*. New York, NY: Blue Ribbon Books (reprinted, 2001, New York, NY: HarperCollins).

Meehan, C. L. (2005). The effects of residential locality on parental and alloparental investment among the Aka foragers of the Central African Republic. *Human Nature* 16: 58-80.

Mesoudi, A. (2011). *Cultural Evolution: How Darwinian Theory Can Explain Human Culture and Synthesize the Social Sciences*. Chicago, IL: University of Chicago Press.

Mesoudi, M. (2016). Cultural evolution: a review of theory, findings and controversies. *Evolutionary Biology* 43: 481-497.

Mesoudi, A. (2017). Pursuing Darwin's curious parallel: prospects for a science of cultural evolution. *Proceedings of the National Academy of Sciences* 114: 7853-7860.

Mesoudi, A. (2021). Cultural selection and biased transformation: two dynamics of cultural evolution. *Philosophical Transactions of the Royal Society B* 376: 20200053.

Mesoudi, A., Chang, L., Dall, S. R. X. and Thornton, A. (2016). The evolution of individual and cultural variation in social learning. *Trends in Ecology and Evolution* 31: 215-225.

Mesoudi, A. and O'Brien, M. J. (2008). The cultural transmission of Great Basin projectile-point technology I: an experimental simulation. *American Antiquity* 73: 3-28.

Mesoudi, A. and Whiten, A. (2004). The hierarchical transformation of event knowledge in human cultural transmission. *Journal of Cognition and Culture* 4: 1-24.

Mesoudi, A., Whiten, A. and Laland, K. N. (2004). Is human cultural evolution Darwinian? Evidence reviewed from the perspective of *The Origin of Species*. *Evolution* 58: 1-11.

Mesoudi, A., Whiten, A. and Laland, K. N. (2006). Towards a unified science of cultural evolution. *Behavioral and Brain Sciences* 29: 329-383.

Minocher, R. and Sommer, V. (2016). Why do mothers harm their babies? Evolutionary Perspectives. *Interdisciplinary Science Reviews* 41: 335-350.

Mitchell, K. J. (2018). *Innate: How the Wiring of Our Brains Shapes Who We Are*. Princeton, NJ: Princeton University Press.

Mithen, S. (1996). *The Prehistory of the Mind*. New York, NY: Thames and Hudson.

Montagu, M. F. A. (Ed.) (1968). *Man and Aggression*. Oxford, UK: Oxford University Press.

Moran, E. F. (1984). Limitations and advances in ecosystems research. In E. F. Moran (Ed.) *The Ecosystem Concept in Anthropology* (pp. 3-32). Washington, DC: AAAS.

Moravec, J. C., Atkinson, Q., Bowern, C., Greenhill, S. J., Jordan, F. M., Ross, R. M., Gray, R., Marsland, S., et al. (2018). Post-marital residence patterns show lineage-specific evolution. *Evolution and Human Behavior* 39: 594-601.

Morgan, C. L. (1896). *Habit and Instinct*. London, UK: Edward Arnold.

Morgan, C. L. (1900). *Animal Behaviour*. London, UK: Edward Arnold.
Morgan, C. L. (1930). *The Animal Mind*. London, UK: Edward Arnold.
Morgan, L. H. (1877). *Ancient Society, or Researches in the Lines of Human Progress From Savagery Through Barbarism to Civilization*. New York, NY: Holt.
Morgan, T. J. H., Rendell, L. E., Ehn, M., Hoppitt, W. and Laland, K. N. (2012). The evolutionary basis of human social learning. *Proceedings of the Royal Society B* 279: 653–662.
Morgan, T. J. H., Uomini, N. T., Rendell, L. E., Chouinard-Thuly, L., Street, S. E., Lewis, H. M., Cross, C. P., Evans, C., et al. (2015). Experimental evidence for the co-evolution of hominin tool-making, teaching and language. *Nature Communications* 6: 6029.
Morin, O. (2013). How portraits turned their eyes upon us: visual preferences and demographic change in cultural evolution. *Evolution and Human Behavior* 34: 222–229.
Morin, O. (2016). *How Traditions Live and Die*. Oxford, UK: Oxford University Press.
Morin, O., Jacquet, P. O., Vaesen, K. and Acerbi, A. (2021). Social information use and social information waste. *Philosophical Transactions of the Royal Society B* 376: 20200052.
Morris, D. (1967). *The Naked Ape*. London, UK: Vintage.
Müller, G. (2017). Why an extended evolutionary synthesis is necessary. *Interface Focus* 7: 20170015.
Muthukrishna, M. (2023). *A Theory of Everyone: The New Science of Who We Are, How We Got Here, and Where We Are Going*. Cambridge, MA: MIT Press.
Muthukrishna, M., Doebeli, M., Chudek, M. and Henrich, J. (2018). The cultural brain hypothesis: how culture drives brain expansion, sociality, and life history. *PLoS Computational Biology* 14: e1006504.
Muthukrishna, M. and Henrich, J. (2019). A problem in theory. *Nature Human Behaviour* 3: 221–229.
Muthukrishna, M., Morgan, T. J. H. and Henrich, J. (2016). The when and who of social learning and conformist transmission. *Evolution and Human Behavior* 37: 10–20.
Narvaez, D., Moore, D. S., Witherington, D. C., Vandiver, T. I. and Lickliter, R. (2022). Evolving evolutionary psychology. *American Psychologist* 77: 424–438.
Navara, K. J. (2018). *Choosing Sexes: Mechanisms and Adaptive Patterns of Sex Allocation in Vertebrates*. Cham, Switzerland: Springer.
Navarrete, A. F., Reader, S. M., Street, S. E., Whiten, A. and Laland, K. N. (2016). The coevolution of innovation and technical intelligence in primates. *Proceedings of the Royal Society B* 371: 20150186.
Nettle, D. (2009). Beyond nature versus culture: cultural variation as an evolved characteristic. *Journal of the Royal Anthropological Institute* 15: 223–240.
Nettle, D. and Frankenhuis, W. E. (2019). The evolution of life-history theory: a bibliometric analysis of an interdisciplinary research area. *Proceedings of the Royal Society B* 286: 20190040.
Nettle, D., Gibson, M. A., Lawson, D. W. and Sear, R. (2013). Human behavioral ecology: current research and future prospects. *Behavioral Ecology* 24: 1031–1040.
Nettle, D. and Scott-Phillips, T. (2023). Is a non-evolutionary psychology possible? In A. du Crest, M. Valković, A. Ariew, H. Desmond, P. Huneman and T. A. C. Reydon (Eds.) *Evolutionary Thinking Across Disciplines: Problems and Perspectives in Generalized Darwinism* (pp. 21–42). Cham, Switzerland: Springer.
Neureiter, N., Ranacher, P., Efrat-Kowlasky, N., Kaiping, G. A., Weibel, R., Widmer, P. and Bouchaert, R. R. (2022). Detecting contact in language trees: a Bayesian phylogenetic model with horizontal transfer. *Humanities and Social Sciences Communications* 9: 205.
Newbury, M. G., Ahern, C. A., Clark, R. and Plotkin, J. B. (2017). Detecting evolutionary forces in language change. *Nature* 551: 223–226.

Norenzayan, A. and Heine, S. J. (2005). Psychological universals: what are they and how can we know? *Psychological Bulletin* 131: 763–784.

Nowak, M. A., McAvoy, A., Allen, B. and Wilson, E. O. (2017). The general form of Hamilton's rule makes no predictions and cannot be tested empirically. *Proceedings of the National Academy of Sciences* 114: 5665–5670.

Nunn, C. L. (1999). The evolution of exaggerated sexual swellings in primates and the graded-signal hypothesis. *Animal Behaviour* 58: 229–246.

Nunn, C. L. (2011). *The Comparative Approach in Evolutionary Anthropology and Biology.* Chicago, IL: University of Chicago Press.

O'Bleness, M., Searles, V. B., Varki, A., Gagneux, P. and Sikela, J. M. (2012). Evolution of genetic and genomic features unique to the human lineage. *Nature Reviews Genetics* 13: 853–866.

Ocobock, C. and Lacy, S. (2024). Woman the hunter: the physiological evidence. *American Anthropologist* 126: 7–18.

Odling-Smee, F. J. (1988). Niche-constructing phenotypes. In H. C. Plotkin (Ed.) *The Role of Behavior in Evolution* (pp. 73–132). Cambridge, MA: MIT Press.

Odling-Smee, J. (2024). *Niche Construction: How Life Contributes to its Own Evolution.* Cambridge, MA: MIT Press.

Odling-Smee, F. J., Laland, K. N., and Feldman, M. W. (1996). Niche construction. *American Naturalist* 147: 641–648.

Odling-Smee, F. J., Laland, K. N. and Feldman, M. W. (2003). *Niche Construction: The Neglected Process in Evolution.* Princeton, NJ: Princeton University Press.

Öhman, A. and Mineka, S. (2001). Fears, phobias, and preparedness: towards an evolved module of fear and fear learning. *Psychological Review* 108: 483–522.

Oldroyd, D. R. (1983). *Darwinian Impacts: An Introduction to the Darwinian Revolution* (2nd edition). Milton Keynes, UK: Open University Press.

Orians, G. H. (1969). On the evolution of mating systems in birds and mammals. *American Naturalist* 103: 589–603.

Otto, S. P., Christiansen, F. B. and Feldman, M. W. (1995). *Genetic and cultural inheritance of continuous traits* (Morrison Institute for Population and Resource Studies Paper No. 64). Stanford CA: Stanford University Press.

Pacheco-Cobos, L., Winterhalder, B., Cuatianquiz-Lima, C., Rosetti, M. F., Hudson, R. and Ross, C. T. (2019). Nahua mushroom gatherers use area-restricted search strategies that conform to marginal value theorem predictions. *Proceedings of the National Academy of Sciences* 116: 10339–10347.

Packer, M. J. and Cole, M. (2019). Evolution and ontogenesis: the deontic niche of human development. *Human Development* 62: 175–211.

Page, A. E. and French, J. C. (2020). Reconstructing prehistoric demography: what role for extant hunter-gatherers? *Evolutionary Anthropology* 29: 332–345.

Pagel, M., Atkinson, Q. D. and Meade, A. (2007). Frequency of word-use predicts rates of lexical evolution throughout Indo-European history. *Nature* 449: 717–720.

Palma, H. A. (2023). An amazing journey: Darwin and the Fuegians. In M. E. B. Pretes (Ed.) *Understanding Evolution in Darwin's 'Origin': The Emerging Context of Evolutionary Thinking* (pp. 59–76). Cham, Switzerland: Springer.

Pampus, J. D. and Daegling, D. J. (2016). The enduring puzzle of the human chin. *Evolutionary Anthropology* 25: 20–35.

Parker, G. A. (1970). The reproductive behavior and the nature of sexual selection in *Scatophaga stercoraria* L. II. The fertilization rate and the spatial and temporal relationships of each sex around the site of mating and oviposition. *Journal of Animal Ecology* 39: 205–228.

Parker, G. A. and Thompson, E. A. (1980). Dung fly struggles: a test of the war of attrition. *Behavioral Ecology and Sociobiology* 7: 37–44.

Perreault, C. (2012). The pace of cultural evolution. *PLoS One* **7**: e45150.

Perrin, N. (1980). *Giving Up the Gun: Japan's Reversion to the Sword, 1543–1879*. Boulder, CO: Shambhala Publications.

Perry, G. H., Dominy, N. J., Claw, K. G., Lee, A. S., Fiegler, H., Redon, R., Werner, J., Villanea, A., et al. (2007). Diet and the evolution of human amylase gene copy number variation. *Nature Genetics* **39**: 1256–1260.

Peyrégne, S., Slon, V. and Kelso, J. (2024). More than a decade of genetic research on the Denisovans. *Nature Reviews Genetics* **25**: 83–103.

Pietschnig, J. and Voracek, M. (2015). One century of global IQ gains: a formal meta-analysis of the Flynn effect (1909–2013). *Perspectives on Psychological Science* **10**: 282–306.

Pigliucci, M. and Müller, G. B. (Eds.) (2010). *Evolution: The Extended Synthesis*. Cambridge, MA: MIT Press.

Pinker, S. (1997). *How the Mind Works*. London, UK: Penguin Books.

Pisor, A. C., Borgerhoff Mulder, M. and Smith, K. M. (2024). Long-distance social relationships can both undercut and promote local natural resource management. *Proceedings of the Royal Society B* **379**: 20220269.

Pollen, A. A., Kilik, U., Lowe, C. B. and Camp, J. G. (2023). Human-specific genetics: new tools to explore the molecular and cellular basis of human evolution. *Nature Reviews Genetics* **24**: 687–711.

Plomin, R. (2018). *Blueprint: How DNA Makes Us Who We Are*. Cambridge, MA: MIT Press.

Plotkin, H. C. (Ed.) (1982). *Learning, Development, and Culture: Essays in Evolutionary Epistemology*. Chichester: Wiley.

Plotkin, H. (1994). *Darwin Machines and the Nature of Knowledge*. London, UK: Penguin Books.

Plotkin, H. (2004). *Evolutionary Thought in Psychology: A Brief History*. Oxford, UK: Blackwell.

Pollet, T. V., Nelissen, M. and Nettle, D. (2009). Lineage based differences in grandparental investment: evidence from a large British cohort study. *Journal of Biosocial Science* **41**: 355–379.

Pollet, T. V., Tybur, J. M., Frankenhuis, W. E. and Rickard, I. J. (2014). What can cross-cultural correlations teach us about human nature? *Human Nature* **25**: 410–429.

Popper, K. R. (1979). *Objective Knowledge: An Evolutionary Approach*. Oxford, UK: Clarendon Press.

Price, G. R. (1970). Selection and covariance. *Nature* **277**: 520–521.

Pritchard, D. J., Tello Ramos, M. C., Muth, F. and Healy, S. D. (2017). Treating hummingbirds as feathered bees: a case of ethological cross-pollination. *Biology Letters* **13**: 20170610.

Prokop, P. and Fancovicova, J. (2016). Mothers are less disgust sensitive than childless females. *Personality and Individual Differences* **96**: 65–69.

Pyke, G. H. (1978). Optimal foraging in hummingbirds: testing the marginal value theorem. *American Zoologist* **18**: 739–752.

Quintana-Murci, L. (2019). Human immunology through the lens of evolutionary genetics. *Cell* **177**: 184–199.

Raby, P. (2001). *Alfred Russel Wallace: A Life*. Princeton, NJ: Princeton University Press.

Raviv, L., Meyer, A. and Lev-Ari, S. (2019). Larger communities create more systematic languages. *Proceedings of the Royal Society B* **286**: 20191262.

Ready, E. and Price, M. H. (2021). Human behavioral ecology and niche construction. *Evolutionary Anthropology* **30**: 71–83.

Rendell, L., Boyd, R., Cownden, D., Enquist, M., Eriksson, K., Feldman, M. W., Fogarty, L., Ghirlanda, S., Lillicrap, T. and Laland, K. N. (2010). Why copy others? Insights from the social learning strategies tournament. *Science* **328**: 208–213.

Rendell, L., Fogarty, L., Hoppitt, W. J. E., Morgan, T. J. H., Webster, M. M. and Laland, K. N. (2011). Cognitive culture: theoretical and empirical insights into social learning strategies. *Trends in Cognitive Sciences* **15**: 68–76.

Rendell, L. E. and Whitehead, H. (2003). Vocal clans in sperm whales (*Physeter macrocephalus*). *Proceedings of the Royal Society B* **270**: 225–231.

Revell, L. J. and Harmon, L. J. (2022). *Phylogenetic Comparative Methods in R*. Princeton, NJ: Princeton University Press.

Rheinberger, H. and Müller-Wille, S. (2018). *The Gene: From Genetics to Postgenomics*. Chicago, IL: University of Chicago Press.

Richards, R. J. (1987). *Darwin and the Emergence of Evolutionary Theories of Mind and Behavior*. Chicago, IL: University of Chicago Press.

Richerson, P. (2018). The use and non-use of the human nature concept by evolutionary biologists. In E. Hannon and T. Lewens (Eds.) *Why We Disagree About Human Nature* (pp. 145–169). Oxford, UK: Oxford University Press.

Richerson, P., Baldini, R., Bell, A. V., Demps, K., Frost, K., Hillis, V., Mathew, S., Newton, E. K., et al. (2016). Cultural group selection plays an essential role in explaining human cooperation: a sketch of the evidence. *Behavioral and Brain Sciences* **39**: e30.

Richerson, P. J. and Boyd, R. (2005). *Not by Genes Alone: How Culture Transformed Human Evolution*. Chicago, IL: Chicago University Press.

Richerson, P. J., Boyd, R. and Henrich, J. (2010). Gene-culture coevolution in the age of genomics. *Proceedings of the National Academy of Sciences* **107**: 8985–8992.

Ridley, M. (2003). *Nature via Nurture: Genes, Experience and What Makes Us Human*. New York, NY: Harper Collins.

Roberts, M. J. (Ed.) (2007). *Integrating the Mind*. New York, NY: Psychology Press.

Roli, A., Jaeger, J. and Kauffman, S. A. (2022). How organisms come to know the world: fundamental limits on artificial general intelligence. *Frontiers in Ecology and Evolution* **9**: 806283.

Romanes, G. J. (1882). *Animal Intelligence*. London, UK: Kegan, Paul, Trench & Co.

Rose, A. C. (2020). *In the Hearts of the Beasts: How American Behavioral Scientists Rediscovered the Emotions of Animals*. New York, NY: Oxford University Press.

Rose, H. and Rose, S. (Eds.) (2000). *Alas Poor Darwin: Arguments Against Evolutionary Psychology*. London, UK: Jonathan Cape.

Rose, S., Lewontin, R. C. and Kamin, L. J. (1984). *Not in Our Genes: Biology, Ideology and Human Nature*. London, UK: Penguin Books.

Rosenbaum, S. and Silk, J. B. (2022). Pathways to parental care in primates. *Evolutionary Anthropology* **31**: 245–262.

Rosenberg, N. A., Pritchard, J. K., Weber, J. L., Cann, H. M., Kidd, K. K., Zhivotovsky, L. A. and Feldman, M. W. (2002). Genetic structure of human populations. *Science* **298**: 2381–2385.

Ross, C. T., Borgerhoff Mulder, M., Oh, S., Bowles, S., Beheim, B., Bunce, J., Caudell, M., Clark, G., et al. (2018). Greater wealth inequality, less polygyny: rethinking the polygyny threshold model. *Journal of the Royal Society Interface* **15**: 20180035.

Royer, C. (1870). *Origine de l'Homme et des Societés*. Paris: Guillaumin.

Rozin, P., Haidt, J. and McCauley, C. R. (2000). Disgust. In: M. Lewis and J. Haviland (Eds.) *Handbook of Emotions* (pp. 637–653). New York, NY: Guilford Press.

Ruiz, A. M. and Santos, L. R. (2013). Understanding differences in the way human and nonhuman primates represent tools: the role of teleological-intentional information. In C. M. Sanz, J. Call and C. Boesch (Eds.) *Tool Use in Animals: Cognition and Ecology* (pp. 119–133). Cambridge, UK: Cambridge University Press.

Ruse, M. (1999). *Mystery of Mysteries: Is Evolution a Social Construct?* Cambridge, MA: Harvard University Press.

Ruse, M. (2017). *Darwinism as Religion: What Literature Tells Us About Evolution*. New York, NY: Oxford University Press.

Salzen, E. A. (1996). Introduction. In: K. Lorenz, *On Aggression* (pp. ix–xxiii). London, UK: Routledge.

Samuelsson, S., Byrne, B., Olson, R. K., Hulslander, J., Wadsworth, S., Corley, R., Willcutt, E. G. and DeFries, J. C. (2008). Response to early literacy instruction in the United States, Australia, and Scandinavia: a behavioral-genetic analysis. *Learning and Individual Differences* **18**: 289–295.

Sánchez-Viallagra, M. R. and van Schaik, C. P. (2019). Evaluating the self-domestication hypothesis of human evolution. *Evolutionary Anthropology* **28**: 133–143.

Schaack, S., Gilbert, C. and Feschotte, C. (2010). Promiscuous DNA: horizontal transfer of transposable elements and why it matters for eukaryotic evolution. *Trends in Ecology and Evolution* **25**: 537–546.

Schaefer, N. K., Shapiro, B. and Green, R. E. (2021). An ancestral recombination graph of human, Neanderthal, and Denisovan genomes. *Science Advances* **7**: eabc0776.

Schecter, S. and Gintis, H. (2016). *Game Theory in Action: An Introduction to Classical and Evolutionary Models*. Princeton, NJ; Princeton University Press.

Schmitt, D. P. (2005). Sociosexuality from Argentina to Zimbabwe: a 48-nation study of sex, culture, and strategies of human mating. *Behavioral and Brain Sciences* **28**: 247–311.

Schneirla, T. C. (1966). Behavioral development and comparative psychology. *Quarterly Review of Biology* **41**: 247–351.

Schniter, E., Kaplan, H. S. and Gurven, M. (2023). Cultural transmission vectors of essential knowledge and skills among Tsimane forager-farmers. *Evolution and Human Behavior* **44**: 530–540.

Scott-Phillips, T., Blancke, S. and Heintz, C. (2018). Four misunderstandings about cultural attraction. *Evolutionary Anthropology* **27**: 162–173.

Scott-Phillips, T. C., Laland, K. N., Shuker, D. M., Dickins, T. E. and West, S. A. (2014). The niche construction perspective: a critical appraisal. *Evolution* **68**: 1231–1243.

Sear, R. (2007). The impact of reproduction on Gambian women: does controlling for phenotypic quality reveal costs of reproduction? *American Journal Physical Anthropology* **132**: 632–641.

Sear, R. (2015). Evolutionary contributions to the study of human fertility. *Population Studies* **69**: S39–S55.

Sear, R. (2016). Beyond the nuclear family: an evolutionary perspective on parenting. *Current Opinion in Psychology* **7**: 98–103.

Sear, R. (2020). Do human 'life history strategies' exist? *Evolution and Human Behavior* **41**: 513–526.

Sear, R. (2021a). Demography and the rise, apparent fall, and resurgence of eugenics. *Population Studies* **75**: S201–S220.

Sear, R. (2021b). The male breadwinner nuclear family is not the 'traditional' human family, and promotion of this myth may have adverse health consequences. *Philosophical Transactions of the Royal Society B* **376**: 20200020.

Sear, R. (2022). 'National IQ' datasets do not provide accurate, unbiased or comparable measures of cognitive ability worldwide. *PsyArXiv*. https://doi.org/10.31234/osf.io/26vfb

Sear, R., Lawson, D. W., Kaplan, H. and Shenk, M. K. (2016). Understanding variation in human fertility: what can we learn from evolutionary demography? *Philosophical Transactions of the Royal Society B* **371**: 20150144.

Sear, R. and Mace, R. (2005). Are humans cooperative breeders? In E. Voland, A. Chasoitis and W. Schiefenhoevel (Eds.) *Grandmotherhood: The Evolutionary Significance of the Second Half of Female Life* (pp. 143–159). Piscataway, NJ: Rutgers University Press.

Sear, R., Mattison, S. M. and Shenk, M. K. (2024). Demography. In J. Koster, B. Scelza and M. K. Shenk (Eds.) *Human Behavioral Ecology* (pp. 307–332). Cambridge, UK: Cambridge University Press.

Seed, A. M., Emery, N. J. and Clayton, N. S. (2009). Intelligence in corvids and apes: a case of convergent evolution? *Ethology* **115**: 401–420.

Segerstråle, U. (1986). Colleagues in conflict: an 'in vivo' analysis of the sociobiology controversy. *Biology and Philosophy* **1**: 53–87.

Segerstråle, U. (2000). *Defenders of the Truth: The Sociobiology Debate*. Oxford, UK: Oxford University Press.

Segestrale, U. (2013). *Nature's Oracle: The Life and Work of W. D. Hamilton*. Oxford, UK: Oxford University Press.

Ségurel, L. and Bon, C. (2017). On the evolution of lactase persistence in humans. *Annual Review of Genomics and Human Genetics* **18**: 297–319.

Shackelford, T. K. (Ed.) (2020). *The SAGE Handbook of Evolutionary Psychology: Integration of Evolutionary Psychology with Other Disciplines*. London, UK: Sage.

Shenk, M. K. (2024). Marriage. In J. Koster, B. Scelza and M. K. Shenk (Eds.) *Human Behavioral Ecology* (pp. 230–255). Cambridge, UK: Cambridge University Press.

Shenk, M. K., Towner, M. C., Kress, H. C. and Alam, N. (2013). A model comparison approach shows stronger support for economic models of fertility decline. *Proceedings of the National Academy of Sciences* **110**: 8045–8050.

Shennan, S. J. and Wilkinson, J. R. (2001). Ceramic style change and neutral evolution: a case study from Neolithic Europe. *American Antiquity* **66**: 577–594.

Shettleworth, S. (2000). Modularity and the evolution of cognition. In C. Heyes and L. Huber (Eds.) *The Evolution of Cognition* (pp. 43–60). Cambridge, MA: MIT Press.

Shipley, B. (2016). *Cause and Correlation in Biology: A User's Guide to Path Analysis, Structural Equations and Causal Inference with R* (2nd edition). Oxford, UK: Oxford University Press.

Shultz, S. and Dunbar, R. I. M. (2007). The evolution of the social brain: anthropoid primates contrast with other vertebrates. *Proceedings of the Royal Society B* **274**: 2429–2436.

Shumaker, R. W., Walkup, K. R. and Beck, B. B. (Eds.) (2011). *Animal Tool Behavior: The Use and Manufacture of Tools by Animals*. Baltimore, MD: Johns Hopkins University Press.

Sibly, R. M. (1983). Optimal group size is unstable. *Animal Behaviour* **31**: 947–948.

Simkin, M. V. and Roychowdhury, V. P. (2003). Read before you cite! *Complex Systems* **14**: 269.

Simpson, G. G. (1944). *Tempo and Mode in Evolution*. New York, NY: Columbia University Press.

Singh, M. and Glowacki, L. (2022). Human social organization during the Late Pleistocene: beyond the nomadic-egalitarian model. *Evolution and Human Behavior* **43**: 418–431.

Sivan, J., Curry, O. S. and Van Lissa, C. J. (2018). Excavating the foundations: cognitive adaptations for multiple moral domains. *Evolutionary Psychological Science* **4**: 408–419.

Skirgård, H., Haynie, H. J., Blasi, D. E., Hammarström, H., Collins, J., Latarche, J. J., Lesage, J., Weber, T., et al. (2023). Grambank reveals the importance of genealogical constraints on linguistic diversity and highlights the impact of language loss. *Science Advances* **9**: eadg6175.

Smiseth, P. T., Wright, J. and Kölliker, M. (2008). Parent-offspring conflict and co-adaptation: behavioural ecology meets quantitative genetics. *Proceedings of the Royal Society B* **275**: 1823–1830.

Smith, E. A. (1985). Inuit foraging groups: some simple models incorporating conflicts of interest, relatedness, and central place sharing. *Ethology and Sociobiology* **6**: 27–47.

Smith, E. A. (1998). Is Tibetan polyandry adaptive? Methodological and metatheoretical analyses. *Human Nature* **9**: 225–261.

Smith, E. A. (2000). Three styles in the evolutionary analysis of human behavior. In L. Cronk, N. Chagnon and W. Irons (Eds.) *Adaptation and Human Behavior: An Anthropological Perspective* (pp. 27–46). New York, NY: Aldine de Gruyter.

Smith, E. A. (2013). Agency and adaptation: new directions in evolutionary anthropology. *Annual Review of Anthropology* **42**: 103–120.

Smith, E. A., Borgerhoff Mulder, M., and Hill, K. (2000). Evolutionary analyses of human behaviour: a commentary on Daly and Wilson. *Animal Behaviour* 60: F21–F26.

Smith, E. A., Borgerhoff Mulder, M., and Hill, K. (2001). Controversies in the evolutionary social sciences: a guide for the perplexed. *Trends in Ecology and Evolution* 16: 128–135.

Smith, S. E. (2020). Is evolutionary psychology possible? *Biological Theory* 15: 39–49.

Sng, O., Neuberg, S. L., Varnum, M. E. W. and Kenrick, D. T. (2018). The behavioral ecology of cultural psychological variation. *Psychological Review* 125: 714–743.

Sniekers, S., Stringer, S., Watanabe, K., Jansen, P. R., Coleman, J. R. I., Krapohl, E., Taskesen, E., Hammerschlag, A. R., et al. (2017). Genome-wide association meta-analysis of 78,308 individuals identifies new loci and genes influencing human intelligence. *Nature Genetics* 49: 1107–1112.

Sober, E. & Wilson, D. S. (1998). *Unto Others: The Evolution and Psychology of Unselfish Behavior*. Cambridge, MA: Harvard University Press.

Soranzo, N., Bufe, B., Sabeti, P. C., Wilson, J. F., Weale, M. E., Marguerie, R., Meyerhof, W. and Goldstein, D. B. (2005). Positive selection on a high-sensitivity allele of the human bitter-taste receptor *TAS2R16*. *Current Biology* 15: 1257–1265.

Spencer, H. (1855). *Principles of Psychology* (1st edition). London, UK: Longman.

Spencer, H. (1857). Progress: its law and cause. *The Westminster Review* 67: 445–485.

Spencer, H. (1870). *Principles of Psychology* (2nd edition). London, UK: Longman.

Sperber, D. (1996). *Explaining Culture: A Naturalistic Approach*. Oxford, UK: Blackwell.

Sperber, D. and Claidière, N. (2006). Why modeling cultural evolution is still such a challenge. *Biological Theory* 1: 20–22.

Sperber, D., Cara, F. and Girotto, V. (1995). Relevance theory explains the selection task. *Cognition* 57: 31–95.

Stark, R. (1997). *The Rise of Christianity: How the Obscure, Marginal Jesus Movement Became the Dominant Religious Force in the Western World in a Few Centuries*. San Francisco, CA: Harper Collins.

Starkweather, K. E. and Hames, R. (2012). A survey of non-classical polyandry. *Human Nature* 23: 149–172.

Starkweather, K. E., Reynolds, A. Z., Zohora, F. and Alam, N. (2023). Shodagor women cooperate across domains of work and childcare to solve an adaptive problem. *Philosophical Transactions of the Royal Society B* 378: 20210433.

Stearns, S. (1992). *The Evolution of Life History*. Oxford, UK: Oxford University Press.

Stearns, S. C., Byars, S. G., Govindaraju, D. R. and Ewbank, D. (2010). Measuring selection in contemporary human populations. *Nature Reviews Genetics* 11: 611–622.

Stedman, H. H., Kozyak, B. W., Nelson, A., Thesier, D. M., Su, L. T., Low, D. W., Bridges, C. R., Shrager, J. B., et al. (2004). Myosin gene mutation correlates with anatomical changes in the human lineage. *Nature* 428: 415–418.

Stephens, D. W. and Krebs, J. R. (1986). *Foraging Theory*. Princeton, NJ: Princeton University Press.

Steward, J. (1955). *Theory of Cultural Change*. Urbana, IL: University of Illinois Press.

Stibbard-Hawkes, D. N. E. (2019). Costly signaling and the handicap principle in hunter-gatherer research: a critical review. *Evolutionary Anthropology* 28: 144–157.

Stott, R. (2012). *Darwin's Ghosts: In Search of the First Evolutionists*. London, UK: Bloomsbury.

Stotz, K. (2014). Extended evolutionary psychology: the importance of transgenerational developmental plasticity. *Frontiers in Psychology* 5: 908.

Stotz, K. and Griffiths, P. (2004). Genes: philosophical analysis put to the test. *History and Philosophy of the Life Sciences* 26: 5–28.

Street, S., Cross, C. P. and Brown, G. R. (2016). Exaggerated sexual swellings in female non-human primates are reliable signals of female fertility and body condition. *Animal Behaviour* 112: 203–212.

Street, S. E., Navarrete, A. F., Reader, S. M. and Laland, K. N. (2017). Coevolution of cultural intelligence, extended life history, sociality, and brain size in primates. *Proceedings of the National Academy of Sciences* 114: 7908–7914.

Stubberfield, J. M. (2022). Content biases in three phases of cultural transmission: a review. *Culture and Evolution* 19: 41–60.

Stulp, G., Sear, R. and Barrett, L. (2016a). The reproductive ecology of industrial societies, part I: why measuring fertility matters. *Human Nature* 27: 422–444.

Stulp, G., Sear, R., Schaffnit, S. B., Mills, M. C. and Barrett, L. (2016b). The reproductive ecology of industrial societies, part II: the association between wealth and fertility. *Human Nature* 27: 445–470.

Sugiyama, M. S. (2017). Oral storytelling as evidence of pedagogy in forager societies. *Frontiers in Psychology* 8: 471.

Sulloway, F. J. (1979). *Freud, Biologist of the Mind: Beyond the Psychoanalytic Legend*. London, UK: Burnett Books.

Sultan, S. E. (2015). *Organism and Environment: Ecological Development, Niche Construction, and Adaptation*. Oxford, UK: Oxford University Press.

Suntsova, M. V. and Buzdin, A. A. (2020). Differences between human and chimpanzee genomes and their implications in gene expression, protein function and biochemical properties of the two species. *BMC Genomics* 21: 535.

Sussman, R. W. (2014). *The Myth of Race: The Troubling Persistence of an Unscientific Idea*. Cambridge, MA: Harvard University Press.

Symons, D. (1987). If we're all Darwinians, what's the fuss about? In C. Crawford, M. Smith, and D. Krebs (Eds.) *Sociobiology and Psychology: Ideas, Issues and Applications* (pp. 121–146). Hillsdale, NJ: Erlbaum.

Symons, D. (1989). A critique of Darwinian anthropology. *Ethology and Sociobiology* 10: 131–144.

Symons, D. (1990). Adaptiveness and adaptation. *Ethology and Sociobiology* 11: 427–444.

Takács, K. (2018). Discounting of evolutionary explanations in sociology textbooks and curricula. *Frontiers in Sociology* 3: 24.

Taylor, S. A. and Larson, E. L. (2019). Insights from genomes into the evolutionary importance and prevalence of hybridization in nature. *Nature Ecology and Evolution* 3: 170–177.

Tebbich, S. and Bshary, R. (2004). Cognitive abilities related to tool use in the woodpecker finch, *Cactospiza pallida*. *Animal Behaviour* 67: 689–697.

Tehrani, J. J. (2013). The phylogeny of Little Red Riding Hood. *PLoS One* 8: e78871.

Tehrani, J. J. and Collard, M. (2002). Investigating cultural evolution through biological phylogenetic analyses of Turkmen textiles. *Journal of Anthropological Archaeology* 21: 443–463.

Tehrani, J., Kendal., J. and Kendal, R. (Eds.) (2023). *The Oxford Handbook of Cultural Evolution*. Oxford, UK: Oxford University Press.

Thorndike, E. L. (1911). *Animal Intelligence*. New York, NY: Macmillan.

Thorpe, W. H. (1979). *The Origins and Rise of Ethology*. London, UK: Heinemann.

Tiger, L. (1969). *Men in Groups*. New York, NY: Random House.

Tiger, L. and Fox, R. (1971). *The Imperial Animal*. New York, NY: Holt, Rinehart, Winston.

Tinbergen, N. (1951). *The Study of Instinct*. Oxford, UK: Oxford University Press.

Tinbergen, N. (1953). *The Herring Gull's World*. London, UK: Collins.

Tinbergen, N. (1963). On aims and methods of ethology. *Zeitschrift fur Tierpsychologie* 20: 410–433.

Tinbergen, N. and Kruyt, W. (1932). Über die orientierung des bienenwolfes (Philanthus Triangulum Fabr.) [On the orientation of the bee wolf]. *Zeitschrift für vergleichende Physiologie* 25: 292–344.

Tomasello, M. (2023). Differences in the social motivations and emotions of humans and other great apes. *Human Nature* 34: 588–604.

Tooby, J. and Cosmides, L. (1989a). Evolutionary psychology and the generation of culture, part I. Theoretical considerations. *Ethology and Sociobiology* 10: 29–49.

Tooby, J. and Cosmides, L. (1989b). Evolutionary psychology and the generation of culture, part II. Case study: a computational theory of social exchange. *Ethology and Sociobiology* 10: 51–97.

Tooby, J. and Cosmides, L. (1990a). The past explains the present: emotional adaptations and the structure of ancestral environments. *Ethology and Sociobiology* 11: 375–424.

Tooby, J. and Cosmides, L. (1990b). On the universality of human nature and the uniqueness of the individual: the role of genetics and adaptation. *Journal of Personality* 58: 17–67.

Tooby, J. and Cosmides, L. (1992). The psychological foundations of culture. In: J. H. Barkow, L. Cosmides and J. Tooby (Eds.) *The Adapted Mind: Evolutionary Psychology and the Generation of Culture* (pp. 137–159). New York, NY: Oxford University Press.

Tooby, J. and Cosmides, L. (2005). Conceptual foundations of evolutionary psychology. In D. Buss (Ed.) *The Handbook of Evolutionary Psychology* (pp. 5–67). Hoboken, NJ: Wiley.

Trivers, R. L. (1971). The evolution of reciprocal altruism. *Quarterly Review of Biology* 46: 35–57.

Trivers, R. L. (1972). Parental investment and sexual selection. In B. Campbell (Ed.) *Sexual Selection and the Descent of Man, 1871–1971* (pp. 136–179). Chicago, IL: Aldine.

Trivers, R. L. (1974). Parent–offspring conflict. *American Zoologist* 14: 249–264.

Trivers, R. L. (1985). *Social Evolution*. Menlo Park, CA: Benjamin Cumins.

Trivers, R. L. and Willard, D. E. (1973). Natural selection of parental ability to vary the sex ratio of offspring. *Science* 179: 90–92.

Turchin, P. (2003). *Historical Dynamics: Why States Rise and Fall*. Princeton, NJ: Princeton University Press.

Turchin, P. (2016). *Ages of Discord: A Structural-Demographic Analysis of American History*. Chaplin, CT: Beresta Books.

Turchin, P., Currie, T. E., Turner, E. A. L. and Gavrilets, S. (2013). War, space, and the evolution of Old World complex societies. *Proceedings of the National Academy of Sciences* 110: 16384–16389.

Turchin, P., Currie, T. E., Whitehouse, H., François, P., Feeney, K., Mullins, D., Hoyer, D., Collins, C., et al. (2018). Quantitative historical analysis uncovers a single dimension of complexity that structures global variation in human social organization. *Proceedings of the National Academy of Sciences* 115: E144–E151.

Turke, P. W. (1990). Which humans behave adaptively, and why does it matter? *Ethology and Sociobiology* 11: 305–339.

Turkheimer, E. (2000). Three laws of behavior genetics and what they mean. *Current Directions in Psychological Science* 9: 160–164.

Tybur, J. M., Lieberman, D. and Griskevicius, V. (2009). Microbes, mating, and morality: individual differences in three functional domains of disgust. *Journal of Personality and Social Psychology* 97: 103–122.

Tylor, E. B. (1865). *Researches into the Early History of Mankind and the Development of Civilization*. London, UK: John Murray.

Tylor, E. B. (1871). *Primitive Culture: Researches into the Development of Mythology, Philosophy, Religion, Art, and Custom, etc.* (Volumes 1 and 2). London, UK: John Murray.

Uchiyama, R., Spicer, R. and Muthukrishna, M. (2022). Cultural evolution of genetic heritability. *Behavioral and Brain Sciences* 45: e152.

Urlacher, S. S. (2023). The energetics of childhood: current knowledge and insights into human variation, evolution, and health. *Yearbook of Biological Anthropology* **181**(Supplement 76): 94–117.

Uyenoyama, M. and Feldman, M. W. (1980). Theories of kin and group selection: a population genetics perspective. *Theoretical Population Biology* **17**: 380–414.

Van de Walle, J., Fay, R., Gaillard, J., Pelletier, F., Hamel, S., Gamelon, M., Barbraud, C., Blanchet, F. G., et al. (2023). Individual life histories: neither slow nor fast, just diverse. *Proceedings of the Royal Society B* **290**: 20230511.

Varki, A., Geschwind, D. H. and Eichler, E. E. (2008). Explaining human uniqueness: genome interactions with environment, behaviour and culture. *Nature Reviews Genetics* **9**: 749–763.

Venkataraman, V. V., Kraft, T. S., Dominy, N. J. and Endicott, K. M. (2017). Hunter-gatherer residential mobility and the marginal value of rainforest patches. *Proceedings of the National Academy of Sciences* **114**: 3097–3102.

Verner, J. and Willson, M. F. (1966). The influence of habitats on mating systems of the North American passerine birds. *Ecology* **47**: 143–147.

Vicedo, M. (2018). The 'disadapted' animal: Niko Tinbergen on human nature and the human predicament. *Journal of the History of Biology* **51**: 191–221.

Voight, B. F., Kudaravalli, S., Wen, X. and Pritchard, J. K. (2006). A map of recent positive selection in the human genome. *PLoS Biology* **4**: e72.

Voland, E. and Stephan, P. (2000). 'The hate that love generated'—sexually selected neglect of one's own offspring in humans. In C. P. van Schaik and C. H. Janson (Eds.) *Infanticide by Males and its Implications* (pp. 447–465). Cambridge, UK: Cambridge University Press.

de Waal, F. (2001). *The Ape and the Sushi Master: Cultural Reflections by a Primatologist*. New York, NY: Basic Books.

Waddington, C. H. (1959). Evolutionary systems: animal and human. *Nature* **183**: 1634–1638.

Wagner, G. P. (Ed.) (2001). *The Character Concept in Evolutionary Biology*. San Diego, CA: Academic Press.

Walker, C. and Thomas, M. G. (2023). Cultural evolution and diet. In J. Tehrani, J. Kendal and R. Kendal (Eds.) *The Oxford Handbook of Cultural Evolution*. Oxford, UK: Oxford University Press. https://doi.org/10.1093/oxfordhb/9780198869252.013.67

Wallace, A. R. (1869). Geological climates and the origin of species. *Quarterly Review* **126**: 359–394.

Walter, K. V., Conroy-Beam, D., Buss, D. M., Asao, K., Sorokowska, A., Sorokowski, P., Aavik, T., Akello, G., et al. (2020). Sex differences in mate preferences across 45 countries: a large-scale replication. *Psychological Science* **31**: 408–423.

Wang, E. T., Kodama, G., Baldi, P. and Moyzis, R. K. (2006). Global landscape of recent inferred Darwinian selection for *Homo sapiens*. *Proceedings of the National Academy of Sciences* **103**: 135–140.

Waring, T. M., Wood, Z. T. and Szathmary, E. (2024). Characteristics processes of human evolution caused the Anthropocene and may obstruct its global solutions. *Proceedings of the Royal Society B* **379**: 20220259.

Wason, P. (1966). Reasoning. In B. M. Foss (Ed.) *New Horizons in Psychology* (pp. 135–151). London, UK: Penguin.

Watson, J. B. (1913). Psychology as the behaviorist views it. *Psychological Review* **20**: 158–177.

Watson, J. B. (1924). *Behaviorism*. New York, NY: Norton.

Wells, J. C. K. and Stock, J. T. (2020). Life history transitions at the origins of agriculture: a model for understanding how niche construction impacts human growth, demography and health. *Frontiers in Endocrinology* **11**: 325.

West-Eberhard, M. J. (2003). *Developmental Plasticity and Evolution*. Oxford, UK: Oxford University Press.

Whitehead, H., Laland, K. N., Rendell, L., Thorogood, R. and Whiten, A. (2019). The reach of gene–culture coevolution in animals. *Nature Communications* **10**: 2405.
Whiten, A., Ayala, F., Feldman, M. W. and Laland, K. N. (2017). The extension of biology through culture. *Proceedings of the National Academy of Sciences* **114**: 7775–7581.
Whiten, A., Goodall, J., McGrew, W. C., Nishida, T., Reynolds, V., Sugiyama, Y., Tutin, C. E. G., Wrangham, R. W. and Boesch, C. (1999). Cultures in chimpanzees. *Nature* **399**: 682–685.
Wilkins, A. S., Wrangham, R. W. and Fitch, W. T. (2014). The 'domestication syndrome' in mammals: a unified explanation based on neural crest cell behavior and genetics. *Genetics* **197**: 795–808.
Wilkinson, G. S. (1984). Reciprocal food sharing in the vampire bat. *Nature* **308**: 181–184.
Williams, G. C. (1966). *Adaptation and Natural Selection: A Critique of Some Current Evolutionary Thought*. Princeton, NJ: Princeton University Press (reprinted 1996).
Williamson, S. H., Hubisz, M. J., Clark, A. G., Payseur, B. A., Bustamante, C. D. and Nielsen, R. (2007). Localizing recent adaptive evolution in the human genome. *PLoS Genetics* **3**: e90.Wilson, D. S. (2002). *Darwin's Cathedral: Evolution, Religion, and the Nature of Society*. Chicago, IL: University of Chicago Press.
Wilson, E. O. (1975a). *Sociobiology: The New Synthesis*. Cambridge, MA: Harvard University Press.
Wilson, E. O. (1975b). Human decency is animal. *The New York Times Magazine* 12 October, 38–50.
Wilson, E. O. (1976). Academic vigilantism and the political significance of sociobiology. *BioScience* **26**: 183–190.
Wilson, E. O. (1978). *On Human Nature*. Cambridge, MA: Harvard University Press.
Wilson, E. O. (1994). *Naturalist*. Washington, DC: Island Press.
Wilson, E. O. (2000). Sociobiology at the end of the century. In E. O. Wilson, *Sociobiology: The New Synthesis* (25th anniversary edition). Cambridge, MA: Harvard University Press.
Winterhalder, B. and Smith, E. A. (2000). Analyzing adaptive strategies: human behavioral ecology at twenty-five. *Evolutionary Anthropology* **9**: 51–72.
Witherington, D. C. and Lickliter, R. (2016). Integrating development and evolution in psychological science: evolutionary developmental psychology, developmental systems, and explanatory pluralism. *Human Development* **59**: 200–234.
Witt, A., Toyokawa, W., Lala, K., Gaissmaier, W. and Wu, C. M. (2023). Social learning with a grain of salt. In M. Goldwater, F. K. Anggoro, B. K. Hayes and D. C. Ong (Eds.) *Proceedings of the 45th Annual Conference of the Cognitive Science Society* (pp. 548–554). Austin, TX: Cognitive Science Society.
Wood, L. A., Vale, G. L., Flynn, E. G. and Rawlings, B. S. (2023). Cross-species comparisons of human and non-human culture: approaches, discoveries, limitations, and future directions. In J. Tehrani, J. Kendal and R. Kendal (Eds.) *The Oxford Handbook of Cultural Evolution*. Oxford, UK: Oxford University Press. https://doi.org/10.1093/oxfordhb/9780198869 252.013.30
Wood, W. and Eagly, A. H. (2012). Biosocial construction of sex differences and similarities in behavior. In J. M. Olson and M. P. Zanna (Eds.) *Advances in Experimental Social Psychology* (Volume 46) (pp. 55–123). Burlington, VT: Academic Press.
Wormley, A. S., Kwon, J. L., Barlev, M. and Varnum, M. E. W. (2023). How much cultural variation around the globe is explained by ecology? *Proceedings of the Royal Society B* **290**: 20230485.
Wrangham, R. (2009). *Catching Fire: How Cooking Made Us Human*. New York, NY: Basic Books.
Wray, G. A., Hoekstra, H. E., Futuyma, D. J., Lenski, R. E., Mackay, T. F. C., Schluter, D. and Strassmann, J. E. (2014). Does evolutionary theory need a rethink? No, all is well. *Nature* **514**: 161–164.

van Wyhe, J. 2005. The descent of words: evolutionary thinking 1780–1880. *Endeavour* 29: 94–100.

Wynne-Edwards, V. (1962). *Animal Dispersion in Relation to Social Behaviour*. Edinburgh, UK: Oliver and Boyd.

Yeh, D. J., Fogarty, L. and Kandler, A. (2019). Cultural linkage: the influence of package transmission on cultural dynamics. *Proceedings of the Royal Society B* 286: 20191951.

Young, A. I., Beonisdottir, S., Przeworski, M. and Kong, A. (2019). Deconstructing the sources of genotype-phenotype associations in humans. *Science* 365: 1396–1400.

Zeder, M. A. (2017). Domestication as a model system for the extended evolutionary synthesis. *Interface Focus* 7: 20160133.

Zeder, M. (2018). Why evolutionary biology needs anthropology: evaluating core assumptions of the extended evolutionary synthesis. *Evolutionary Anthropology* 27: 276–284.

Zefferman, M. R. (2023). Greater variance in male intelligence test scores can be explained by sex-biased test design. *OSF Preprints*. https://doi.org/10.31219/osf.io/83wma

Zefferman, M. R. and Mathew, S. (2015). An evolutionary theory of large-scale human warfare: group-structured cultural selection. *Evolutionary Anthropology* 24: 50–61.

Zeitsch, B. P. and Sidari, M. J. (2020). A critique of life history approaches to human trait covariation. *Evolution and Human Behavior* 41: 527–535.

Zeller, E., Timmermann, A., Yun, K., Raia, P., Stein, K. and Ruan, J. (2023). Human adaptation to diverse biomes over the past 3 million years. *Science* 380: 604–608.

Zhang, H., Ji, T., Pagel, M. and Mace, R. (2020). Dated phylogeny suggests early Neolithic origin of Sino-Tibetan languages. *Scientific Reports* 10: 20792.

Zhen, Y., Huber, C. D., Davies, R. W. and Lohmueller, K. E. (2021). Greater strength of selection and higher proportion of beneficial amino acid changing mutations in humans compared with mice and *Drosophila melanogaster*. *Genome Research* 31: 110–120.

Zhou, Y., Song, H. and Ming, G. (2024). Genetics of human brain development. *Nature Reviews Genetics* 25: 26–45.

Zug, R. and Uller, T. (2022). Evolution and dysfunction of human cognitive and social traits: a transcriptional regulation perspective. *Evolutionary Human Sciences* 4: e43.

Zuk, M. (2002). *Sexual Selections: What We Can and Can't Learn About Sex From Animals*. Berkeley, CA: University of California Press.

Zuk, M. and Simmons, L. W. (2018). *Sexual Selection: A Very Short Introduction*. Oxford, UK: Oxford University Press.

Index

For the benefit of digital users, indexed terms that span two pages (e.g., 52–53) may, on occasion, appear on only one of those pages.

Tables and figures are indicated by an italic *t* and *f* following the page number.

Ache, Paraguay 9, 91–92, 101, 103–4, 208–9
acquired characteristics, inheritance of 26–27, 166–67
adaptation, definition of 95–96, 96*f*
adaptationism 132–33
adaptationist 96–97, 132–33, 223–24
adaptive, definition of 95–96, 96*f*
adaptive lag 101, 109, 115, 131–32, 212*t*, *see also* mismatch
adaptive tradeoffs 81, 84–86, 91–93, 201
Agassiz, L. 31
aggression 1, 3, 45–47, 51, 52–53, 61, 64, 72, 112, 123–24, 184–85, 221–22
Aka, Central African Republic 94–95
Alexander, R. D. 71–72, 79, 98
altruism 54–57, 70, 156–57
  reciprocal 13–14, 52, 53–54, 60–61, 76, 117, 118–19
animal behaviour, comparisons with human behaviour 7–9, 203–6
Aoki, K. 181
Ardrey, R. 47, 73–74
asocial learning 150
associationism 21
associative learning 31*t*, 134–35

Baldwin, J. M. 31–32, 35
Barkow, J. H. 110
Barrett, H. C. 116
Bateson, P. 4, 43
BaYaka, Congo Basin 213
behavioural genetics 160, 179–80, 185–88
behaviourism 36–37, 42, 109–10, 218–19
Benedict, R. 37–38
biogenetic law 28–30, 34
Blackmore, S. 141
Blurton Jones, N. 91–92
Boas, F. 37–38
Borgerhoff Mulder, M. 5–6, 90, 98, 99–100

Bowlby, J. 46, 109, 113–14, 129–30
Boyd, R. 142, 148, 151–52, 164–65, 170, 183–84
brain evolution 102, 191–94, 205–6
Brigham, C. 36
Brodie, R. 141
Brown Blackwell, A. 74–75
Brown, D. 112–13
Buss, D. 110–12, 115, 120–22

Caldwell, C. 152
capitalism 12, 17–18, 28
Carnegie, A. 28
Cavalli-Sforza, L. 142, 145–46, 169–70, 181
Chagnon, N. 79
Charnov, E. 79
cheater detection 117–19
chimpanzees 8, 20, 38–39, 74–75, 93, 94–95, 143–44, 191–93, 194, 204–6
cognitive attractor 162, 163–65
cognitive gadgets 195–97
comparative psychology 33, 42–43, 44
conformity 148, 150–51, 183–84, 202–3
  definition of 150–51
content biases 148, 160–61, 163–64, 167–68
context biases 148, 161, 163–64, 167–68, 212*t*, 217
context-dependent strategies 112–13, 122, 149, 215–16
cooking 132–33, 173–74, 219
Coolidge, C. 36
cooperation 28, 76, 90–91, 94–95, 156–57, 158, 170–71, 174–75, 182–85, 197, 206
cooperative childrearing 86–87, 93–95
Cosmides, L. 5–6, 97, 108, 110, 111–12, 114, 118–19, 128, 129, 130, 133–34, 135, 218
Crook, J. 79, 89
Crook, S. 89
cultural attraction theory 162–65

cultural drift 146*t*, 149, 155
cultural evolution 5, 15, 77, 199–200, 203, 207–8, 210, 212*t*, 216, 217, 219
   case studies 149–59
   critical evaluation 159–67
   cumulative 150–53, 161, 196
   history of 139–42
   hypothesis generation and testing 211–13
   and infanticide 209–10
   key concepts 142–49
   parallels with genetic evolution 144–47, 165–67
   speed of 171–74
   summary of 202–3
Cultural Evolution Society 142
cultural group selection 156–57, 179–80, 182–85, 197–98
cultural niche construction 175, 187, 195, 197–98, 219
cultural selection 15, 142, 145–47, 146*t*, 163–65, 171, 177, 178, 179, 183–84
   definition of 145–46
cultural transmission, biases in 148, *see also* content biases; direct biases; frequency-dependent biases; guided variation; indirect biases; success biases
culture 142–44, 189–91
   alternative views on 215–17
culture–gene coevolution *see* gene–culture coevolution
culturgens 67
Curtis, V. 125
Cuvier, G. 26–27

dairy farming 15–16, 179–82
dairy products 160, 180–81, 190
Daly, M. 5–6, 122–24, 179, 209
Darwin, C. 1, 6, 12–13, 17–23, 34–35, 43, 46, 47–49, 50, 54, 57, 74–75, 119–20, 124–25, 132, 139–40, 144–45, 153, 155, 200–1, 221
   on human behaviour 18–23
   against progressive evolution 26–32
Dawkins, R. 5–6, 13–14, 51, 53–54, 57, 63–64, 68, 71–72, 76, 77, 132, 140–41, 142
Deloria, E. C. 37
demographic transition 86–87, 92–93, 215
Denisovans 192
Dennett, D. 139, 140, 141, 165
developmental bias 221–23

DeVore, I. 74, 79–80, 108
Dickemann, M. 59–60
digger wasps 40–41
direct biases 148, 163–64, 202–3
Ditfurt, Germany 208–9
Dobzhansky, T. 38
domain-general 115–16, 133–36, 161, 197, 216, 217
domain-specific 111, 133–34, 161, 163–64, 217, *see also* domain specificity
domain specificity 115–16, 127, 134, 136, 137
domestication 158, 173–75, 180–81, 182, 185, 220–22, *see also* self-domestication
Drummond, H. 28
dual-inheritance theory 5–6, 170, *see also* gene–culture coevolution
dung flies 62–63
Dunn, M. 155–56

Eibl-Eibesfeldt, I. 46
Ekman, P. 46
Endler, J. 132
environment of evolutionary adaptedness (EEA) 109, 111, 113–15, 120, 127, 128–30, 137
epigenetic inheritance 26–27, 221–22
epigenetic rules 148
ethology 12–13, 33, 39–44, *see also* human ethology
eugenics 12, 17–18, 23–25, 28, 35–36, 38, 65, 213–14
evolutionarily stable strategy (ESS) 52, 62
evolutionary demography 78, 100, 215
evolutionary developmental psychology 113
evolutionary epistemology 139
evolutionary game theory 13–14, 62–64, 76
evolutionary psychology 5, 14–15, 77, 106–7, 191, 193, 199–200, 207–8, 210, 212*t*, 218, 220
   case studies 116–27
   critical evaluation 127–36
   and culture 195–96, 215–17
   history of 108–11
   hypothesis generation and testing 211–13, 214
   and infanticide 209
   key concepts 111–16
   summary of 202
evolved psychological mechanisms, description of 111–13

exaptation 96, 96f, 102–3, 133
  definition of 96
expendable male hypothesis 94–95
extended evolutionary synthesis (EES) 200, 214–15, 220–23
extended phenotype 195–96

facial expressions 20, 46, 124–25
facial resemblance 209
fast-slow life histories 206–7, 213–14
fathers 6–7, 26–27, 56, 94–95, 208–9, 213
Feldman, M. 142, 145–46, 169–70, 181, 187–88
fertility 91–92, 105–6, 209–10, 215
Firmin, A. 31–32
Fisher, R. A. 38, 54–55
foraging 78, 80–81, 87–88, 89, 95, 103, 110, 113–14, 125–26, 134–35, 144, 182, 195–96, 200, 205–7, 213, 215, *see also* optimal foraging theory
Fox, R. 47
FOXP2 gene 192–93
frequency-dependent biases 148
Freud, S. 12–13, 34
Frisch, K. von 40, 44, 50

Gabbra, Kenya 92
Galton, F. 25, 12, 23–25
Garcia, J. 116
gene–culture coevolution 5, 15–16, 77, 159, 167–68, 199–200, 207–8, 212t, 216, 217, 219
  case studies 179–88
  constructing models of 176–79
  critical evaluation 179–97
  history of 169–71
  hypothesis generation and testing 211–13
  and infanticide 209–10
  key concepts 171–79
  summary of 203
gene's-eye view 50–51, 52–54, 57–58, 76
genetic determinism 9–12, 49, 52–54, 68–70, 127, 200–1
Gould, S. J. 3, 4, 65, 66, 96, 165–66
grandmother hypothesis 94
grandmothers 80–81, 94–95
Gray, R. 154–55
group selection 52–55, 56–57, 68, 79–80, 132, 148, 156–57, *see also* cultural group selection

Guglielmino, C. R. 144
guided variation 148, 163–64, 166–67

Hadza, Tanzania 103–4
Haeckel, E. 28–30, 37–38
Haldane, J. B. S. 38, 54–55, 79
Hall, G. S. 34, 35
Hamilton, W. D. 3, 4–5, 13–14, 50–51, 54–57, 64, 68, 108, 110
haplodiploidy 56
Hawkes, K. 79, 101
Heinroth, O. 39–40
Henrich, J. 164–65, 183–85
heritability 179–80, 185–88
  definition of 186
Heyes, C. 196, 197
Hill, K. 5–6, 91–92, 101
Hinde, R. 40, 46, 79
Hitler, A. 29
Holden, C. 104
Holocene 131, 173
homicide 116–17, 122–24
homosexual 7–8, 47, 57, 73
Howell, N. 91
Hrdy, S. B. 74–75, 77, 208
Human Behaviour and Evolution Society (HBES) 4–5
human behavioural ecology 5, 14, 77, 199–200, 207–8
  case studies 86–95
  critical evaluation 95–106
  and culture 215
  history of 78–81
  hypothesis generation and testing 211–13, 212t, 214
  and infanticide 208–9
  key concepts 81–86
  summary of 201
human ethology 44–47, 49, 73–74
human nature 37–38, 46t, 64, 112–13, 200, 214–15, 217–20, 223
human sociobiology 5–6, 13–14, 23, 36, 49, 51, 80, 108, 127, 139–40, 169–70, 223–24
  critical evaluation of 68–76
  and infanticide 208
  summary of 200–1
  see also human sociobiology debate; sociobiology
human sociobiology debate 64–68, 200–1, 203–4

Hume, D. 21
hummingbirds 83–84
hunter-gatherer 80–81, 93, 100–1, 109, 114, 115, 128–29, 154, 174–75, 182–83
Hurtado, A. M. 91–92
Huxley, J. 38, 43–44
Huxley, T. 19, 31–32, 33, 47
hydraulic model 41–42
Hymenoptera 30*t*, 54, 56–57

imprinting 33, 39–40, 49
inclusive fitness 55–57, 94
indirect biases 148, 151–52, 202–3
infanticide 59–60, 74–75, 122–23, 200, 206–7, 208–10, 223
innate 42–43, 44, 47, 108, 110, 155–56, 161, 218–19
instinct 8–9, 12–13, 20, 30*t*, 32–34, 35–36, 37, 38, 39–44, 45, 46*t*, 183, 218–19
intelligence 15–16, 20–21, 27, 29–30, 30*t*, 36, 45, 65, 72–73, 96, 120, 133–34, 170–71, 179–80, 185–86, 205–6
  heritability of 187–88
  quotient (IQ) 187–88
inter-birth interval 86, 91–92
Inuit, Canada 9, 24–25
Irons, W. 79

James, H 34–35
James, W. 34–35
jealousy 112
Jordan, F. 154–55

Kameda, T. 150
Kamin, L. J. 68, 69, 72
Kant, I. 21
Kelvin, W. T., Baron 26
kin selection 13–14, 52, 53–57, 61, 73, 76, 183
Kipsigis, Kenya 90
Kirby, S. 152–53
!Kung San, Kalahari 79–80, 91

Lack, D. 53, 85–86
lactase persistence 180–82, 197–98
Lamarck, J.-B. de 17–18, 26–27, 28–29, 34
Lamarckian 27, 35, 38, 48, 166–67
language 96, 104, 112–13, 115, 136, 139, 140, 142, 144–45, 146*t*, 149–50, 152–56, 165–66, 170–71, 183–84, 190, 192–93, 196–97, 205, 218

Lashley, K. S. 40
Law of Effect 135
learning *see* asocial learning; associative learning; social learning
Lee, R. 79–80, 91
Lehrman, D. 42–44
Lenin, V. I. 37
Lewontin, R. 3, 4, 65, 66–67, 68–70, 72, 132–33, 187–88, 220–21
Li, N. 209–10
Locke, J. 21, 23
Lorenz, K. 12–13, 39–42, 43–47, 50, 52–53
Lubbock, J. 30–31
Lumsden, C. 67–68, 77, 148, 169, 170
Lynch, A. 141

Mace, R. 92, 101, 104, 106
Machery, E. 218
Malthus, T. 18, 20
marriage 1, 25, 86–87, 89–91, 103, 104, 142, 195–96, 208–9
Marx, K. 27, 144–45
Marxist 3, 65, 66–67
mate choice 103, 115, *see also* mate preference; mating
mate preference 119–22, *see also* mate choice; mating
mating 21–22, 43–44, 52–53, 62–63, 80–81, 84–85, 88, 101, 116–17, 123–24, 144, 179, *see also* mate choice; mate preference
Maynard Smith, J. 3, 4, 13–14, 17, 50–51, 53, 55, 62–63, 64, 68, 71, 98–99
Mayr, E. 38
McDougall, W. 35–36
Mead, M. 37–38
memes 5–6, 77, 139–42, 146*t*, 159–60, 165, 200–1
meme's-eye view 200–1
memetics 5–6, 139–40, 141–42, 159–60, 162–63, 165
Mendel, G. J. 38
mental ability 12, 19, 20–21, 22–25, 29–30, 31, 33
Mesoudi, A. 151–52, 162–63
Mill, J. S. 21
Millen, A. 152
mismatch 101, 109, 115, 131–32, 202, 215–16, *see also* adaptive lag
modern synthesis (MS) 38, 221
Morgan, C. L. 33–34

Morgan, L. H. 30–31, 31*t*
Morgan, T. 151
Morgan's canon 33–34
Morin, O. 162, 164
Morris, D. 12–13, 46–47, 73–74
Morton S. G. 31
mothers 6–7, 20, 33, 34, 46, 56, 57, 58–59, 74–75, 91–92, 93–95, 113–14, 125–27, 129–30, 206–7, 208–9, 213
MYH16 gene 173–74

Nakanishi, D. 151–52
nature-nurture debate 32–39, 44, 71
Nazism 4, 17–18, 44–45
Neanderthals 170–71, 173, 192–93
neoteny 132–33
niche construction 171, 174–76, 187–88, 195, 197–98, 203, 207–8, 220–23, *see also* cultural niche construction

O'Brien, M. 151–52
optimal foraging theory 83–84, 86–88, 96–97
ovulation, concealed 8

Pagel, M. 155
parent-offspring conflict 57–60
Parker, G. 62–63
Pavlov, I. P. 37
phenotypic correlation 85–86, 91–92, 105–6
phenotypic gambit 98–99, 215
phylogenetic methods 142, 149–50, 153–56, 158–59, 166, 205–6, 217
Piaget, J. 35
Pinker, S. 110
plasticity-led evolution 221–23
Pleistocene 109, 110, 114, 115, 128–29, 130–32, 136–37, 173, 202, 212*t*
    definition of 109
Plotkin, H. 139
polyandry 89–90, 97, 144–45
polygyny 90–91
polygyny threshold model 90–91
Price, G. 56–57, 62
prosociality 179–80, 182–85

racism 17–18, 31, 38–39, 44–45, 48, 49, 71, 194, 203, *see also* eugenics; Nazism
reaction norms 98, 100, 106–7, 111–12, 135–36, 220, 222–23
reciprocal altruism *see* altruism

reductionism 68–70, 175–76
religion 3, 51, 72, 103–4, 142, 144–46, 157
reproductive ecology 99–100
Rescorla-Wagner rule 134–35
Richerson, P. 142, 148, 151–52, 170
Rockefeller, J. D. 28
Romanes, G. 29–30, 30*t*, 33
Rose, S. 68, 69, 71–72
Royer, C. 74–75

Santa Barbara school 108, 137
Schneirla, T. C. 42–43
self-domestication 185, *see also* domestication
Seville Statement on Violence 45–46, 46*t*
sexual selection 12–13, 21–22, 34, 58, 119–20, 123–24, 179
Shodagor, Bangladesh 94–95
Simpson, G. 38
Smith, E. A. 5–6, 87–88, 98–99, 103, 210
Social Darwinism 12, 17, 28, 48
social learning 15–16, 100, 126–27, 143–44, 150–53, 161, 166–67, 183–84, 195, 196, 197–98, 204–6, 213, 216, *see also* social learning strategies
social learning strategies 147, 149, 161, *see also* social learning
sociobiology 3–5, 13–14, 50–51
    key concepts 52–64
    rejection by social scientists 75–76
    see also human sociobiology
Sociobiology Study Group 65
Spalding, D. 32, 33, 39–40
Spencer, H. 17–18, 27–30, 34–35, 158–59
Sperber, D. 162–63
success biases 164
Sumner, W. 28
Symons, D. 96–97, 108

Thorndike, E. 135
Thorpe, W. H. 40
Tibet 36, 89
Tiger, L. 47
Tinbergen, E. 46
Tinbergen, N. 6–7, 40–41, 43–45, 46, 50, 91, 97–98, 200, 207–8, 210, 222–23
Tooby, J. 5–6, 97, 108, 110, 111–12, 114, 128, 129, 130, 133–34, 135, 218
tool use 8–10, 20, 30*t*, 31*t*, 128, 143–44, 146–47, 152, 172–73, 195, 204–5, 206, 218
transmission chains 152–53, 162–63

Trivers, R. 3, 4, 13–14, 50–51, 57–61, 64, 68, 74, 80, 108, 119–20, 132
Trivers–Willard hypothesis 59–60
Tsimane, Bolivia 103–4, 145, 213
Turchin, P. 157–97
Turner, C. H. 39–40
twins, study of 12, 15–16, 24, 185–86
Tylor, E. 30–31

vampire bats 61
Vrba, E. 96

Wallace, A. 20
Washburn, S. 79–80
Wason, P. 117, 118–19
Wason selection test 117, 119
Watson, J. 36, 37
whales 9–10, 87, 205
Wheeler, W. M. 40

Whiten, A. 162–63
Whitman, C. O. 39–40
Wilkinson, G. 61
Willard, D. 24–25, 60
Williams, G. C. 50–51, 53, 64, 68, 96, 108, 132–33
Wilson, E. O. 3–4, 5–6, 13–14, 18, 50–51, 52, 57–58, 68, 74, 75, 76–77, 108, 148, 218
  criticism of 64–68, 80
  and gene–culture coevolution 169–70
  and genetic determinism 68–70
  and prejudice 70–72
  and storytelling 63–74
Wilson, M. 5–6, 122–24, 179, 209
Wright, R. 127
Wright, S. 38
Wynne-Edwards, V. C. 52–54

Yerkes, R. 36